Understanding our Valuable Tenants: Gut Bacteria

Understanding our Valuable Tenants: Gut Bacteria

Shivsanjeevi Sripathi
and Prerna Pandey

Delve
PUBLISHING
www.delvepublishing.com

Our gut bacteria and we go hand in hand: together in sickness and health!

Please note that this book is not meant for diagnostic or therapeutic purposes. Any disease should be treated under the guidance and supervision of a relevant medical expert.

Chapter 1

Introducing the Gut Microbiota: The Heroes of Our Book

"If there's one thing to know about the human body; it's this: the human body has a ringmaster. This ringmaster controls your digestion, your immunity, your brain, your weight, your health and even your happiness. This ringmaster is the gut"

— Nancy Mure

Shiv Sanjeevi Sripathi[1]

[1]Shiv Sanjeevi Sripathi secured second rank in Mumbai University and first rank in his college at his undergrad level and in his college during his Masters and has worked on a stem cell project at the Specialized Centre for Cell Based Therapy (SCCT), KEM Hospital at Mumbai, following which he was a Junior Research Fellow at Microbiology & Cell Biology Department, Indian Institute of Science, Bangalore from August 2009 to November 2011. The field of work involved cloning of cell wall genes and transcription factors in E. coli & M. smegmatis. This was followed by a stint as a Research and Development Executive at Transasia Biomedicals, Mumbai where he worked on patent writing and Immunoassays from 2011 till 2013. In 2013, he entered the teaching field where he has been tutoring biology and chemistry to the students of 11th and 12th grades along with English grammar and communication skills to people of all age groups. Shiv also loves helping people write scientific manuscripts and has co-edited several books on life sciences and aspects of biotechnology with Apple Academic Press.

Accounting for 250–400 m^{2}, the human gastrointestinal (GI) tract accounts is one of the largest interfaces among the host and environmental factors. 60 tons of food pass through the human GI tract across the average lifetime of an individual. The **gut microbiota** is the bacteria, archaea and eukarya that have colonized the GI tract. This is an association that has developed across evolution and is described as mutual and intricate. Estimates point out to these microbes reaching a quantity of 10^{14} while other studies point out to a 1:1 ratio of human: bacterial cells (Thursby and Juge, 2017).

The initial 4-36 months of our life is crucial in the development of the native and core microbiota. This involves a slew of factors ranging from the delivery type, the gestational age, the type of milk feeding, the duration of weaning along with the cultural practices (inclusive of the diet). The stability of the gut microbiota composition is recorded when a human child is two to three years of age. The diversity and richness portray what is termed as a healthy gut (Rinninella et al, 2019).

This symbiotic association of the gut microbiota with us is often illustrated as the bacteria and humans forming a ***"superorganism"***((Thursby and Juge, 2017; Rinninella et al, 2019).

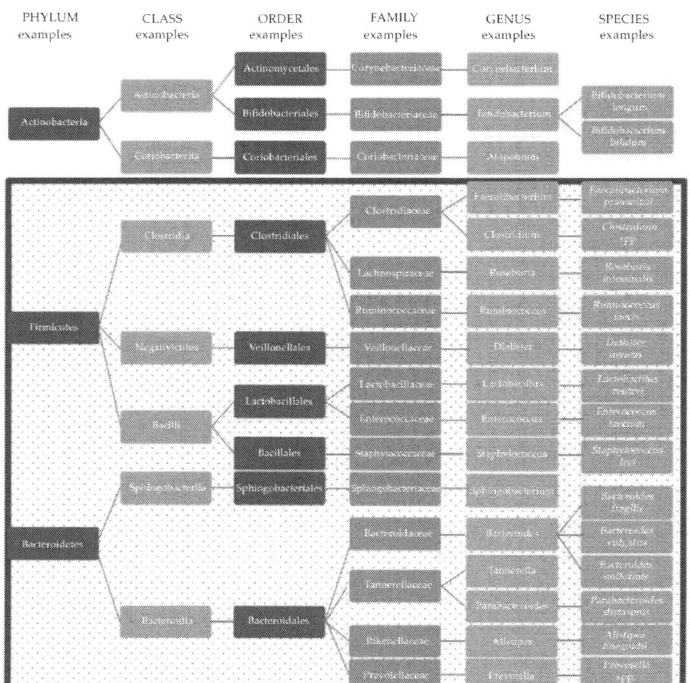

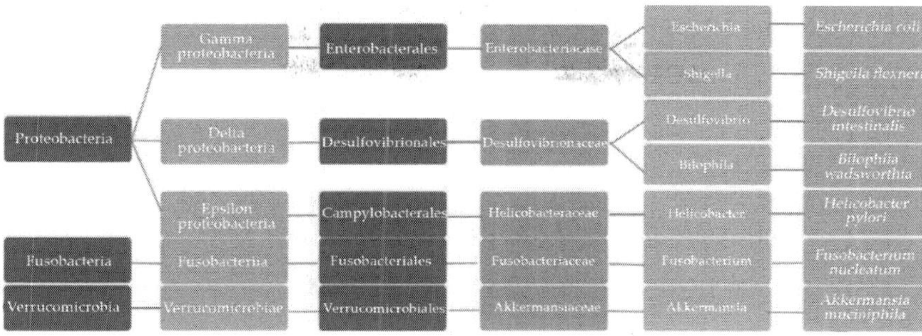

Figure 1: illustrative of the taxonomic gut microbiota composition (Rinninella et al, 2019: *Licensee MDPI, Basel, Switzerland. This article is an open-access article distributed under the terms and conditions of the Creative Commons Attribution (CC BY) license (http://creativecommons.org/licenses/by/4.0/): You are free to share — copy and redistribute the material in any medium or format, Adapt — remix, transform, and build upon the material for any purpose, even commercially*).

90% of gut microbiota is constituted by the bacterial phyla Firmicutes and Bacteroidetes. The former is majorly inclusive of *Clostridium* followed by *Lactobacillus, Ruminicoccus, Bacillus, and Enterococcus. The latter is represented by Bacteroides* and *Prevotella. The Bifidobacterium* genus of Actinobacteria along with the phyla Proteobacteria, Fusobacteria, and Verrucomicrobia are the other residents of our gut (Rinninella et al, 2019).

Fungi account for 3,300-fold lesser diversity than bacteria. While an exact consensus is lacking with *Candida* as the major genus followed by *Saccharomyces*, and *Malassezia. Methanobrevibacter is the most ubiquitous genus of Archaea followed by Methanosphaera*, *Nitrososphaera*, *Thermogynomonas*, and *Thermoplasma. If we look at viruses, eukaryotic viruses account for less than 10% with the remaining chiefly constituted by phages that infect bacteria (*Hillman et al, 2017).

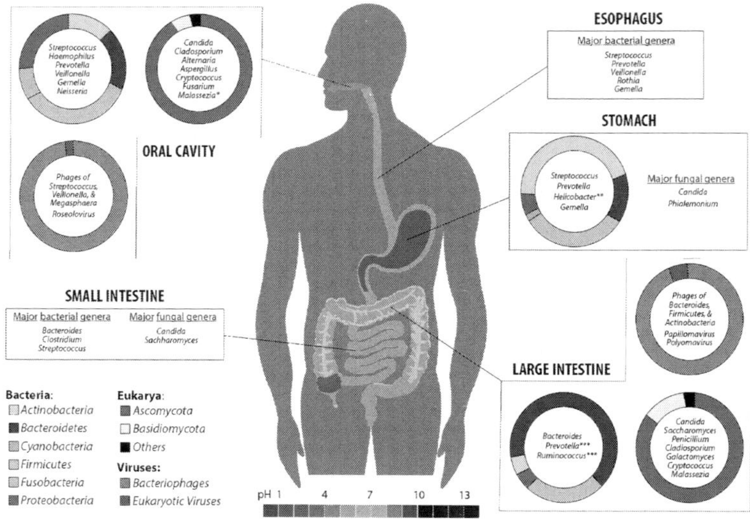

Figure 2: Above is demonstrative of bacteria and viruses in our gastrointestinal (GI) tract on the basis of their location and pH scale (* Malassezia shows conflicting results in studies. ** Helicobacter abundance displays a great variation. *** Based on diet, age and geographical location) (Hillman et al, 2017 by Japanese Society of Microbial Ecology / Japanese Society of Soil Microbiology / Taiwan Society of Microbial Ecology / Japanese Society of Plant Microbe Interactions: *This is an open-access article distributed under the terms of the Creative Commons Attribution License, which permits unrestricted use, distribution, and reproduction in any medium, provided the original work is properly cited*).

Further, there is a variation in the gut microbiota attributed to age, antibiotic use and "growing up".

Table 1: The variations in the microbiota across individuals: * *Unknown genera* (Rinninella et al, 2019: *Licensee MDPI, Basel, Switzerland. This article is an open access article distributed under the terms and conditions of the Creative Commons Attribution (CC BY) license (http://creativecommons.org/licenses/by/4.0/): You are free to: Share — copy and redistribute the material in any medium or format, Adapt — remix, transform, and build upon the material for any purpose, even commercially*)

		Gut Microbiota Abundance							**Bacteria Diversity**
		Actinobacteria	**Bacteroidetes**	**Firmicutes**	**Proteobacteria**	**Fusobacteria**	**Verrucomicrobia**	**Euryarchaeota**	
Anatomical part of gut tract	**Small intestine**			*Lactobacillus*	*Enterobacteriaceae* *				
	Colon		*Bacteroidaceae* * *Prevotellaceae* * *Rikenellaceae* *	*Lachnospiraceae* * *Ruminococcaceae* *					
Gestational age	**Preterm birth (<37 weeks of gestation)**	*Bifidobacterium* spp.↓ *Atopobium* spp.↓	Bacteroides *↓ (non-secretor mothers)	Firmicutes *↓ (non-secretor mothers) *Lactobacillus*↑ *Ruminococcus* spp. *Lachnospiraceae* * *Peptostreptococcaceae* * *Clostridiaceae* *	*Enterobacteriaceae* *↑ *Enterococcus* spp.↑				↓
	Full-term birth	*Bifidobacterium* spp.↑	Bacteroidetes *↑	*Ruminococcus* spp. *Lachnospiraceae* * *Peptostreptococcaceae* * *Clostridiaceae* *	*Enterobacteriaceae* *				↑
Type of delivery	**Vaginal delivery**	*Bifidobacterium* spp.↑ *Bifidobacterium longum*↑ *Bifidobacterium catenulatum*↑	*Prevotella*↑ *Bacteroides fragilis*↑	*Lactobacillus*↑ *Staphylococcus*↑ *Streptococcus* ↑	*Escherichia*↑	*Sneathia*↑			↑
	C-section	*Corynebacterium*↑ *Propionibacterium*↑	*Bacteroides* *↓	*Staphylococcus*↑	*Escherichia*↓ *Shigella*↓				↓

Methods of (milk) feedings	**Breast milk**	*Bifidobacterium*↑↑		*Lactobacillus*↑ *Staphylococcus* ↑	*Enterococcus*↑				↑
	Artificial milk	*Bifidobacterium*↑	*Bacteroides*↑	*Clostridium*↑ *Clostridium difficile*↑ *Lactobacillus*↑	*Escherichia*↑				↓
	Introduction of solid food	*Bifidobacterium*↑	Bacteroidetes *↑ *Bacteroides*↑	Firmicutes *↑ Lactobacilli↑ *Clostridium coccoides*↑					↑
Human age	**Childhood (first year of life)**	*Bifidobacterium*	*Bacteroides*	*Veillonella* *C. coccoides* *C. botulinum*			*Akkermansia muciniphila*		↑
	2–3 years old to adult	*Bifidobacteriaceae* * *Coriobacteriaceae* *	*Bacteroidaceae* * *Prevotellaceae* * *Rikenellaceae* *	*Lachnospiraceae* *Ruminococcaceae*	Proteobacteria *	*Fusobacteria* *	*Akkermansia muciniphila*	*Methanobrevibacter smithii*	↑
	Over 70	*Bifidobacteriaceae*↓		*Clostridium* *↓	Proteobacteria *↑				↓
Antibiotic treatments	**Macrolide**	Actinobacteria *↓	*Bacteroides*↑	Firmicutes *↓	Proteobacteria *↑				↓
	Clarithromycin	Actinobacteria *↓	*Bacteroides*↑	Firmicutes *↓	Proteobacteria *↑				↓
	Vancomycin			*Lactobacillus*↓ *Clostridium*↓					↓
	Ciprofloxacin	*Bifidobacterium*↓	*Alistipes*↓ Bacteroides↑	*Faecalibacterium*↓ *Oscillospira*↓ *Ruminococcus*↓ *Dialister*↓					↓
	Clindamycin	*Bifidobacteriaceae*↓ *Lactobacillus*↓							

Table 2: The variations in the abundance of gut microbiota across individuals based on the enterotypes and body mass index (BMI) level along with extrinsic factors inclusive of diet, ethnicity, cultural practices, exercise and lifestyle: * *Unknown genera. (1) Main genera* (Rinninella et al, 2019: *Licensee MDPI, Basel, Switzerland. This article is an open access article distributed under the terms and conditions of the Creative Commons Attribution (CC BY) license (http://creativecommons.org/licenses/by/4.0/): You are free to: Share — copy and redistribute the material in any medium or format, Adapt — remix, transform, and build upon the material for any purpose, even commercially*)

		Gut Microbiota Abundance							**Bacteria Diversity**	**Ref.**
		Actinobacteria	**Bacteroidetes**	**Firmicutes**	**Proteobacteria**	**Verrucomicrobia**	**Euryarchaeota**	**Spirochaetes**		
Enterotype	**I**	*Slackia*	*Bacteroides* (1) *Parabacteroides*	*Clostridiales* * *Alkaliphilus* *Lactobacillus* *Catenibacterium*	*Geobacter*		*Methanobrevibacter smithii*			[13]
	II	*Eggerthella*	*Prevotella* (1)	*Veillonella* *Ruminococcaceae* * *Holdemania* *Peptostreptococcaceae* * *Staphylococcus* *Leuconostoc*	*Desulfovibrionaceae* * *Rhodospirillum* *Helicobacter* *Escherichia* *Shigella*	*Akkermansia muciniphila*				
	III	*Gordonibacter*	*Sphingobacterium*	*Ruminococcus*(1) *Staphylococcus* *Marvinbryantia* *Symbiobacterium* *Ruminococcaceae* * *Dialister*		*Akkermansia muciniphila*				
BMI level (among children and adolescents)	**Overweight to obese level**	*Bifidobacterium*↑ *B. vulgatus*↓	Bacteroidetes *↓	*Lactobacillus*↑ Firmicutes *↑	Proteobacteria *↑				↓	[46,47,49,50]
	Underweight level			*Roseburia* spp.↓ Ruminococcus↓ Clostridium spp.↓	*Enterobacteriaceae* *↑		*Methanobrevibacter smithii*↑		↓	[46,51]

Exercise frequency	**High performance sport**		*Bacteroides*↓	*Ruminococcaceae* *↑ *Lactobacillus*↓ *Lactobacillaceae* *↓	*Succinivibrionaceae* *↑	*Akkermansiaceae*↓			↑	[57]
	Children and adolescents daily exercise			*Clostridiales*↑ *Roseburia*↑ *Lachnospiraceae*↑ *Erysipelotrichaceae*↑					↑	[46]
Type of diet, communities, climate, and geographical sites	**High-fiber diet (African diet)**	*Bifidobacterium*↓	*Prevotella*↑ *Bacteroides*↓	*Eubacterium*↑ *Oscillibacter*↑ *Butyricicoccus*↑ *Sporobacter*↑ *Blautia*↓ *Dorea*↓ *Lachnospiraceae* *↓ *Roseburia*↓ *Faecalibacterium*↓ *Ruminococcus*↓ *Erysipelotrichaceae* *↓	*Succinivibrio*↑			*Treponema*↑	↑	[53,55]
	High-fat and animal protein diet (Western diet)	Actinobacteria *↑	*Bacteroides*↑ *Alistipes*↑ *Barnesiella*↑	*Roseburia*↓ *Eubacterium rectale*↓ *Ruminococcus bromii*↓	*Bilophila*↑				↓	

Looking at the functions of the gut microbiota, as of now, they include:

- They function to regulate digestion in the GI tract
- They produce and absorb short-chain fatty acids (SCFAs), amino acids, vitamins, bile acids and lipids.
- They exert a protective effect from pathogens and circumvent the colonization by the latter.
- They contribute to the maintenance of the intestinal epithelium.
- They also influence and regulate the working of the immune cells (innate and adaptive).

(Thursby and Juge, 2017; Rinninella et al, 2019)

Metabolites	Functions	References
Short-chain fatty acids (SCFAs): E.g., Acetate, butyrate, propionate, hexanoate, valerate	• Regulate host metabolic pathways via G-protein-coupled receptor GPR41 or GPR43 -mediated signaling: • energy homeostasis; synthesis of glucagon-like peptide 1 (GLP-1); increase leptin production. • Improve glucose tolerance and insulin sensitivity. • Potent histone deacetylase (HDAC) inhibitor - regulation of intestinal cell proliferation. • Intestinal gluconeogenesis, lipogenesis, suppression of fasting-induced adipose factor Fiaf (lipoprotein lipase inhibitor) in intestinal epithelium. • Immunomodulatory effect, activate dendritic cells, gut immunity.	Blottière et al., 2003; Backhed et al., 2004; Xiong et al., 2004; Samuel et al., 2008; Tolhurst et al., 2012; De Vadder et al., 2014
Indole derivatives: E.g., Indole, indoxyl sulfate, indole-3-propionic acid (IPA)	• IPA as powerful antioxidant, inhibitor of amyloid-beta fibril formation, and exhibits neuroprotective and cytoprotective effects against a variety of oxidotoxins. • IPA regulates intestinal barrier function via the xenobiotic sensor, pregnane X receptor (PXR), in which it reduces intestinal inflammation (downregulates enterocyte pro-inflammatory cytokines TNF-α), and regulate intestinal permeability and mucosal integrity (upregulates junctional protein-coding mRNAs). • Indoxyl sulfate as uremic toxin that accumulates in the blood of patients with impaired excretion system.	Bendheim et al., 2002; Deguchi et al., 2002; Wikoff et al., 2009; Venkatesh et al., 2014
Bile acid metabolites: E.g., Deoxycholic acid (DCA), lithocholic acid (LCA)	• Activate host nuclear receptors and cell signaling pathways: regulation of bile acid, cholesterol, glucose, lipid, and energy metabolism. • Exhibit antimicrobial effects.	Inagaki et al., 2006; Hylemon et al., 2009
Choline metabolites: E.g., Choline, trimethylamine N-oxide (TMAO) and betaine	• Modulate lipid metabolism and glucose homeostasis. • Contribute to non-alcoholic fatty liver disease and cardiovascular disease.	Dumas et al., 2006. Wang Z. et al., 2011
Phenolic derivatives:	• Antimicrobial effects: repress pathogenic microbes, influence gut microbiota composition, maintenance of intestinal health.	Larrosa et al., 2006, 2010; Lee H.C. et al., 2006; Selma et al., 2009

E.g., 4-OH phenylacetic acid, equol, urolithins, enterolactone, enterodiol, 8-prenylnaringenin, 2-(3,4-dihydroxyphenyl)acetic acid, 3-(4-hydroxyphenyl)propionic acid, 5-(3,4-dihydroxyphenyl)valeric acid	• Protective effect against oxidative stress. • Estrogen-modulating effect. • Platelet aggregation inhibition effect. • Urolithin exhibits anti-inflammatory and cancer chemopreventive effects.	
Vitamins:	• Energy production, red blood cell formation, as enzymatic cofactor for diverse biochemical reactions.	LeBlanc et al., 2011; Nicholson et al., 2012; Forster et al., 2017; Lerner et al., 2017b
E.g., Thiamine (B1), riboflavin (B2), niacin (B3), pyridoxine (B6), pantothenic acid (B5), biotin (B7), folate (B11–B9), cobalamin (B12), and menaquinone (K2)	• DNA replication, repair and methylation, regulating cell proliferation.	
	• Production of nucleotides, vitamins and amino acids. • Enhance immune functioning.	
Polyamines:	• Sustain high proliferation rate of intestinal epithelial cells.	Guo et al., 2003, 2005; Perez-Cano et al., 2010; Johnson et al., 2015; Rooks and Garrett, 2016
E.g., putrescine, spermidine, and spermine	• Dysregulated polyamine metabolism possibly enhances cancer development. • Enhance intestinal barrier integrity and function via stimulating synthesis of intercellular junction proteins • [occludin, zonula occludens-1 (ZO-1), E-cadherin]. • Enhance maturation of intestinal and systemic adaptive immune system. • Spermine inhibits pro-inflammatory M1 macrophage activation.	

Figure 3: Above summarizes the roles played by the gut microbiota in terms of the metabolites they produce (Copyright © 2018 Kho and Lal. *This is an open-access article distributed under the terms of the Creative Commons Attribution License (CC BY). The use, distribution or reproduction in other forums is permitted, provided the original author(s) and the copyright owner(s) are credited and that the original publication in this journal is cited, in accordance with accepted academic practice. No use, distribution or reproduction is permitted which does not comply with these terms*).

Let us see an interesting aspect of the gut microbiota as outlined by research.

Case Study: Gut Microbes Influence the Behavior

Table 3: Is our "personality" shaped by genetics and our environment? A slew of psychiatric or behavioral issues has been associated with changes in the gut microbiota. (Katerina V.-A. Johnson, 2020: *This is an open-access article under the CC BY license: http://creativecommons.org/licenses/BY/4.0/*):

Genus	Change in abundance	Behavioral trait/psychiatric condition	Study subject	References
Akkermansia	↓	Autism	Children	[52]
	↑	Autism	Children	[53]
	↓	Stress	Mice	[54], [55]

Alistipes	↓	Autism	Children	[56]
	↑	Stress	Mice	[57]
	↑	Depression	Adults	[58], [59]
Bacteroides	↑	Autism	Children	[53], [60]
	↓	Stress	Mice	[55], [61]
	↑	Negative mood	Adults	[62]
	↑	Psychosis	Adults	[63]
Bifidobacterium	↓	Autism	Children	[52], [53], [60], [64], [65], [66], [67]
	↑	Autism	Children	[68]
	↓	Autism	Mice	[69]
	↓	Stress	Mice	[70], [71]
	↑	Stress	Mice	[72]
	↑	Negative mood	Adults	[62]
	↓	Depression	Adults	[73]
Blautia	↓	Autism	Infants	[74]
	↓	Autism	Children	[75]
	↓	Autism	Mice	[69]
	↓	Schizophrenia	Adults	[76]
Clostridium	↑	Autism	Children	[53], [56], [67], [75], [77], [78], [79], [80]
	↓	Autism	Children	[60]
	↑	Stress	Mice	[61], [72], [81]
	↑	Depression	Adults	[59], [82]
	↑	Schizophrenia	Adults	[76]
Collinsella	↑	Autism	Children	[56]
	↓	Autism	Children	[60]
	↑	Schizophrenia	Adults	[76]
Corynebacterium	↑	Autism	Children	[56]
	↓	Stress	Rats	[83]
Desulfovibrio	↑	Autism	Children	[60], [80]
Dialister	↓	Autism	Children	[56], [60]
	↑	Sociability	Infants	[84]
	↓	Depression	Adults	[18], [59], [85]
Dorea	↑	Autism	Children	[53], [56]
	↓	Autism	Children	[75]
	↓	Stress	Mice	[61]

Faecalibacterium	↑	Autism	Infants	[74]
	↓	Autism	Children	[53], [86]
	↓	Negative mood	Adults	[87]
	↑	Negative mood	Adults	[62]
	↓	Depression	Adults	[59], [85]
	↑	Depression (prenatal)	Adults	[88]
	↓	Bipolar disorder	Adults	[89], [90]
Flavonifractor	↑	Autism	Children	[75]
	↑	Depression	Adults	[59]
Lactobacillus	↑	Autism	Children	[56], [64], [68], [80], [91]
	↓	Autism	Children	[53]
	↓	Stress	Mice	[61], [72], [92]
	↓	Stress	Rats	[83]
	↓	Stress	Macaques	[93]
	↓	Depression	Adults	[73]
	↑	Psychosis	Adults	[63]
Lactococcus	↓	Autism	Children	[53], [60]
Oscillibacter	↑	Depression	Adults	[58]
	↑	Stress	Rats	[83]
Oscillospira	↓	Autism	Children	[53]
	↑	Sociability	Mice	[94]
	↓	Stress	Mice	[55], [94], [95]
Parabacteroides	↓	Autism	Children	[56]
	↑	Autism	Children	[60]
	↑	Sociability	Infants	[84]
	↓	Stress	Mice	[55], [61], [92], [94]
	↑	Negative mood	Adults	[62]
	↑	Depression	Adults	[59]
Prevotella	↑	Autism	Children	[53]
	↓	Autism	Children	[91]
	↑	Stress	Mice	[94]
	↓	Negative mood	Adults	[62]
	↑	Depression	Adults	[82]
	↓	Depression	Adults	[18]

Roseburia	↑	Autism	Children	[53]
	↑	Stress	Mice	[61]
	↓	Stress	Mice	[55]
	↓	Negative mood	Adults	[62]
	↓	Schizophrenia	Adults	[76]
Streptococcus	↓	Autism	Children	[53], [60]
	↑	Depression	Adults	[82]
Sutterella	↑	Autism	Children	[52], [65], [96], [97]
	↓	Autism	Children	[75]
	↑	Stress	Mice	[55]
Turicibacter	↓	Autism	Children	[53]
	↑	Sociability	Mice	[94]
	↓	Stress	Mice	[70], [72], [94]
	↑	Depression	Adults	[18]

Overall, stress and anxiety were overall associated with alterations in the gut microbial composition and lower diversity. Interestingly, the author states" *In conclusion, differences in gut microbiome composition and diversity are shown to be linked to personality traits in the general population. The results of this study add a new dimension to our understanding of personality and are in line with accumulating evidence that the gut microbiome can influence the central nervous system in humans, with effects on behavior*" (Katerina V.-A. Johnson, 2020).

Isn't this exciting?

To present other diseases where the gut microbiome is altered, refer to Table 4 below (Copyright © 2018 Kho and Lal. *This is an open-access article distributed under the terms of the Creative Commons Attribution License (CC BY). The use, distribution or reproduction in other forums is permitted, provided the original author(s) and the copyright owner(s) are credited and that the original publication in this journal is cited, in accordance with accepted academic practice. No use, distribution or reproduction is permitted which does not comply with these terms*):

Disease categories	Specific diseases	Associated dysbiotic features	Reference
Immune-mediated/autoimmune diseases	inflammatory bowel disease (IBD)	• Increase in virulent gut microbes (*Enterobacteriaceae* species, *Bacteroides fragilis*) and mucolytic *Ruminococcus* sp. • Decrease in butyrate-producing *Firmicutes* (such as *Faecalibacterium prausnitzii*, *Roseburia hominis*)	Sokol et al., 2008; Png et al., 2010; Willing et al., 2010; Machiels et al., 2014
	Irritable bowel syndrome (IBS)	• Increase in *Escherichia coli* • Decrease in *Clostridium leptum* group of bacteria and *Bifidobacterium*. • Decrease in bile acid biotransformation	Duboc et al., 2012
	Celiac disease	• Increase in *Bacteroides–Prevotella* group • Decrease in *Bifidobacterium* • Varying observation (decrease or no change) in *Clostridium histolyticum*, *C. lituseburense*, and *Faecalibacterium prausnitzii* • Alteration in SCFAs composition, but overall increase in total SCFA	Tjellström et al., 2005; Nadal et al., 2007; De Palma et al., 2010
	Systemic lupus erythematosus (SLE)	• Increase in *Blautia* sp. and Gram-negative bacteria such as *Protecbacteria*. • Decrease in gut microbiota diversity, *Odoribacter* sp., *Alistipes* sp. • Increase in serum endotoxin	Shi et al., 2014; Luo et al., 2018
	Type-1 diabetes	• Increase in *Bacteroidetes* • Decrease in *Actinobacteria*, *Firmicutes*, and *Firmicutes/Bacteroidetes* ratio	Murri et al., 2013
	Rheumatoid arthritis (RA)	• Increase in *Prevotella copri* and decrease in *Bacteroides* sp. in new-onset RA • Increase in microbiota diversity of *Lactobacillus* genus in early RA	Liu et al., 2013; Scher et al., 2013
	Atopic disease (E.g., childhood allergic asthma)	• Increase in fecal burden of *Clostridium difficile*, and *C. difficile/Bifidobacteria* ratio	Kalliomäki et al., 2001
Metabolic disorders/cardiovascular disorders	Obesity	• Increase in *Firmicutes*, *Actinobacter* • Varying observation (decrease, no change, increase) in *Bacteroidetes* • Increase in glycoside hydrolase and SCFAs (butyrate and acetate)	Turnbaugh et al., 2006, 2009a; Kollada et al., 2017
	Type-2 diabetes	• Increase in *Lactobacillus* • Decrease in *Clostridium coccoides*, *Atopobium* cluster, and *Prevotella* • Decrease in butyrate biosynthesis	Qin et al., 2012; Sato et al., 2014
	Hypertension	• Increase in the Firmicutes/Bacteroidetes ratio, lactate-producer • Decrease in microbiota diversity, acetate- and butyrate-producers	Yang et al., 2015
	Atherosclerosis	• Increase in metabolites TMAO, endotoxin level (risk factor for early atherosclerosis)	Wiedermann et al., 1999; Koeth et al., 2013
Cancer	Colorectal cancer (CRC)	• Increase in enterotoxigenic *Bacteroides fragilis*, and pathobionts *Fusobacterium* and *Campylobacter* sp. • Decrease in butyrate-producer (*Faecalibacterium* and *Roseburia*)	Wang et al., 2012; Wu et al., 2013
Neurop sy chi atri c	Autism spectrum disorder (ASD)	• Increase in *Clostridium* sp., *Bacteroidetes*, *Lactobacillus*, *Desulfovibrio* • Decrease in *Bifidobacteria*	Song et al., 2004; Adams et al., 2011
	Alzheimer's disease	• Possible connection between gut microbiota-synthesized amyloids, LPS, γ-aminobutyric acid (GAB A – major inhibitory neurotransmitter), and the increased permeability of gut barrier and blood brain barrier with age	Pistollato et al., 2016
	Depression	• Increase in genus *Eggerthella*, *Holdemania*, *Gelria*, *Turicibacter*, *Paraprevotella*, *Anaerofilum* • Decrease in gut microbiota diversity, *Prevotela* and *Dialister*	Kelly et al., 2016
	Parkinson's Disease	• Increase in anti-inflammatory butyrate-producers from genus *Blautia*, *Coprococcus*, and *Roseburia* in patient fecal sample, pro-inflammatory *Proteobacteria* in patient mucosa • Increased gene expression in LPS biosynthesis and microbial type III secretion system	Keshavarzian et al., 2015
Infectious disease	*Clostridium difficile* infection (CDI)	• Increase in *Clostridium difficile*; • Decrease in gut microbiota diversity and secondary bile acids-producing *Clostridium scindens*	Theriot et al., 2014
Uremic disease	Chronic kidney disease	• Increase in *Firmicutes*, *Proteobacteria*, and *Actinobacteria* • Decrease in *Lactobacilli*	Vaziri et al., 2013

Let us go ahead with the other chapters that discuss the gut microbiota in disease and health.

REFERENCES:

1. Thursby E, Juge N. Introduction to the human gut microbiota. *Biochem J*. 2017;474(11):1823-1836. Published 2017 May 16. doi:10.1042/BCJ20160510
2. Rinninella E, Raoul P, Cintoni M, et al. What is the Healthy Gut Microbiota Composition? A Changing Ecosystem across Age, Environment, Diet, and Diseases. *Microorganisms*. 2019;7(1):14. Published 2019 Jan 10. doi:10.3390/microorganisms7010014
3. Hillman, E. T., Lu, H., Yao, T., & Nakatsu, C. H. (2017). Microbial Ecology along the Gastrointestinal Tract. *Microbes and environments*, *32*(4), 300–313. https://doi.org/10.1264/jsme2.ME17017
4. Kho ZY and Lal SK (2018) The Human Gut Microbiome – A Potential Controller of Wellness and Disease. *Front. Microbiol.* 9:1835. doi: 10.3389/fmicb.2018.01835
5. Katerina V.-A. Johnson (2020). Gut microbiome composition and diversity are related to human personality traits. Human Microbiome Journal (15): 100069: https://doi.org/10.1016/j.humic.2019.100069.

Chapter 2

The Gut Bacteria, Lung Diseases and COVID-19

Shiv Sanjeevi Sripathi[1]

[1]Shiv Sanjeevi Sripathi secured second rank in Mumbai University and first rank in his college at his undergrad level and in his college during his Masters and has worked on a stem cell project at the Specialized Centre for Cell Based Therapy (SCCT), KEM Hospital at Mumbai, following which he was a Junior Research Fellow at Microbiology & Cell Biology Department, Indian Institute of Science, Bangalore from August 2009 to November 2011. The field of work involved cloning of cell wall genes and transcription factors in E. coli & M. smegmatis. This was followed by a stint as a Research and Development Executive at Transasia Biomedicals, Mumbai where he worked on patent writing and Immunoassays from 2011 till 2013. In 2013, he entered the teaching field where he has been tutoring biology and chemistry to the students of 11th and 12th grades along with English grammar and communication skills to people of all age groups. Shiv also loves helping people write scientific manuscripts and has co-edited several books on life sciences and aspects of biotechnology with Apple Academic Press.

2.1 THE GUT AND THE LUNG

Research is uncovering what is called the **gut-lung axis** that involves communication between the microbes of the gut and the lungs. The gut shows dominance of Bacteroidetes and Firmicutes while Bacteroidetes, Firmicutes, and Proteobacteria are seen in the lung. The phyla are however reported to be along the same lines in both these organs while the species are evidently different. The gut lung axis facilitates a channel for the passing of hormones, microbial metabolites, and endotoxins into the bloodstream that connects the gut niche and the lung. This axis is bidirectional with alterations induced in the gut microbiota and the blood when there is an inflammation in the lungs (Zhang et al, 2020).

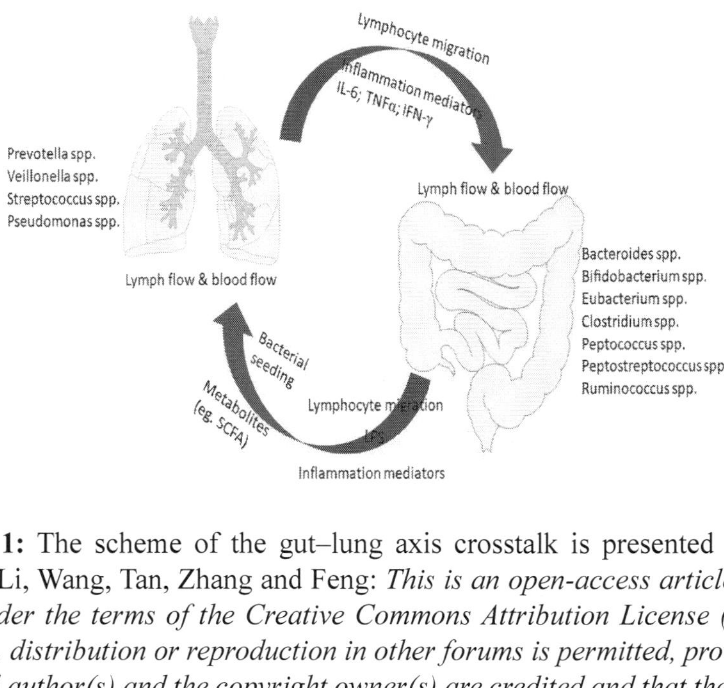

Figure 1: The scheme of the gut–lung axis crosstalk is presented in (2020 Zhang, Li, Wang, Tan, Zhang and Feng: *This is an open-access article distributed under the terms of the Creative Commons Attribution License (CC BY). The use, distribution or reproduction in other forums is permitted, provided the original author(s) and the copyright owner(s) are credited and that the original publication in this journal is cited, in accordance with accepted academic practice. No use, distribution or reproduction is permitted which does not comply with these terms*).

The role of gut microbiota has been extended to the realm of several lung diseases of which a snapshot is presented below.

2.2 THE GUT AND ASTHMA

Ranking as a major and ubiquitous non-communicable disease, asthma ranks 28th as a leading cause of burden while it occupies the 16th spot for among the leading causes of years lived with disability (Dharmage et al, 2019). The global number of individuals with asthma is pitched at 300 million with 100 million more predicted to be affected (The Global Asthma Report, 2018).

The condition (childhood or adult-onset) has similar causes with the interaction between genetics and environmental triggers accounting for the condition.

Table 1: below lists asthma triggers in the environment between childhood and adulthood: *BD, bronchodilator; COPD, chronic obstructive pulmonary disease; HMW, high molecular weight;* FEV_1*, forced expiratory volume in 1 second; LMW, low molecular weight;* O_3*, ozone; PM, particulate matter.*

†*A web-based list of asthmagens can be found at www.occupationalasthma.com.*

(Dharmage, Perret and Custovic, 2019: *This is an open-access article distributed under the terms of the Creative Commons Attribution License (CC BY). The use, distribution or reproduction in other forums is permitted, provided the original author(s) and the copyright owner(s) are credited and that the original publication in this journal is cited, in accordance with accepted academic practice. No use, distribution or reproduction is permitted which does not comply with these terms*):

Environmental exposure	Feature	Children	Adoles-cents	Adults
ALLERGIC				
Airborne triggers - House dust mite (HDM) - Animal hair and dander - Pollen exposure - Mold (fungal) spores - Thunderstorm asthma	HDM aeroallergen is a perennial asthma trigger linked to asthma incidence in high-risk children	(94–96)	(94)	(94)
	Association between pet allergen exposure and asthma is conflicting and non-conclusive	(94, 97, 98)	(98)	(98)
	Grass pollen triggers asthma exacerbations requiring emergency department attendances	(99)	(99)	
	Indoor fungal spore exposure can worsen asthma control; decreased visible indoor mold reduces symptoms but not PEFM variability	(100, 101)	(100, 101)	(100, 101)
	High outdoor Alternaria exposure may contribute to severe asthma/ respiratory arrest (ages 11–25)	(102)	(103)	(103)
	Fungal spores, especially Cladosporium are associated with asthma hospitalization (ages 2–17)	(104)	(104)	
	IgE sensitization to mycoses linked to neutrophilic airway inflammation and lower lung function			(105)
	Thunderstorm asthma can be triggered by outdoor pollen, and possibly fungal, spores	(106, 107)	(106, 107)	(106, 107)
Food allergens (n = 170), e.g.,	In asthma, c/w non-atopy, odds for current asthma increased 3.8-fold if food allergy was likely	(108)	(108)	(108)
- Egg white -Peanut	Food allergy is an uncommon trigger in asthma, but may present as life-threatening asthma, especially to peanut and other tree nut allergens	(109, 110)		
- Tree nuts - Shellfish - Cows milk	Co-existing, poorly-controlled asthma is a risk factor for severe or fatal food-induced anaphylaxis	(110, 111)		

Occupational sensitizing agents (with latency)	Extensive lists of occupational asthmagens known to cause new-onset occupational asthma and/or exacerbate pre-existing asthma† - Sensitizing HMW agents (e.g., plant allergens like flour, flowers, latex; animal allergens by animal handlers and lab workers; biological enzymes; fungi-yeasts) - Sensitizing LMW agents (e.g., chemicals like isocyanates, reactive dyes, industrial cleaning/sterilizing agents; metals; pharmaceuticals like antibiotics, opiates; solder flux; wood dusts)		(112–115)	(112–115)
NON-ALLERGIC				
Non-allergic triggers				
- Respiratory viral infections - Cold air - Humidity - Exercise	Typically trigger variations in asthma symptoms and airflow limitation, and their presence increases the probability that the individual has asthma	(94)	(94)	(94)
Tobacco smoke exposure-Parental smoking	Parental smoking linked to increased incidence of childhood asthma and wheeze	(116)		
- Second-hand smoke exposure	Incident asthma from regular smoking from late childhood; may be greater for those non-allergic compared with allergic, and for those exposed to maternal smoking *in utero*	(117, 118)	(117, 118)	
- Personal smoking	Personal smoking may worsen asthma control/ exacerbations		(119)	(119)
	Personal smoking predisposes to post-BD airflow obstruction and asthma-COPD overlap			(120)
Traffic-related air pollution (TrAP)	Air pollutants (O_3, $PM_{2.5}$) linked to new asthma cases and increased ED admissions globally	(121)	(121)	(121)
-Car exhaust fumes	Long-term exposure to PM_1 may worsen asthma, especially for young males with childhood allergy	(122)	(122)	
	Pollution-related decline in FEV_1 growth was similar between those with and without asthma	(123, 124)	(123, 124)	
	Residential TrAP may contribute to new-onset and persistence of asthma in middle-aged adults			(125)
	Natural experiment of diesel exhaust exposure and adverse short-term changes in spirometry			(126)

Household air pollution (HAP)	Non-polluting home heating improved asthma symptoms, days off school, and healthcare utilization	(127)		
Occupational agents (no latency)	Airway irritants (e.g., chlorine, ammonia) - Irritant occupational asthma - Work exacerbated (pre-existing) asthma		(113, 114)	(113, 114)

Research published in 2011explored the impact of the diversity of the intestinal microbiota (neonatal fecal flora) on asthma. The gut microbes were scored in a birth cohort of 411 high-risk children as part of the Copenhagen Prospective Study on Asthma in Childhood. Traditional culturing and 16S rRNA PCR analyses were done of the flora were done at 1 and 12 months ages. Assessments were conducted for 6 years. The authors reported an inverse association between bacterial diversity in the early intestinal flora 1 and 12 months following birth and the risk of allergic rhinitis and peripheral blood eosinophils while this was not observed for asthma or atopic dermatitis sowing the impact of intestinal microbiota on the allergic disease (Bisgaard et al, 2011).

In a 2017-published work, a mouse asthma model was studied for how microbial exposures impact gut colonization to look at whether asthma can be prevented. Assignment of mice was done into 4 groups:

A: offsprings were kept in a specific pathogen-free (SPF) environment during fetal, lactation, and childhood periods;

B, offsprings were kept in the SPF environment during fetal and lactation periods, and kept in the general environment during childhood;

C, offsprings were kept in the SPF environment only during the fetal period, and then kept in the general environment; and

D, offsprings were kept in the general environment the whole time.

Denaturing gradient gel electrophoresis was employed to assess the intestinal flora diversity. A mice model of asthma was established using OVA followed by appropriate analyses.

The diversity of intestinal microflora was evidently more in group D (general environment the whole time) against groups A, B and C at three days and three weeks following birth while the diversity of intestinal microflora was conspicuously more in groups C and D against groups A and B at five weeks following birth. Thus, animals reared in a general environment had more intestinal diversity than those in an SPF environment.

Table 2: The pathological scores for the asthmatic mice were lower in the animals reared in the general environment as against those maintained in the SPF surroundings, as shown in (Qian et al, 2017: *Open-Access: This article is distributed under the terms of the Creative Commons Attribution 4.0 International License (http://creativecommons.org/licenses/by/4.0/), which permits unrestricted use, distribution, and reproduction in any medium, provided you give appropriate credit to the original author(s) and the source, provide a link to the Creative Commons license, and indicate if changes were made. The Creative Commons Public Domain Dedication waiver (http://creativecommons.org/publicdomain/zero/1.0/) applies to the data made available in this article, unless otherwise stated*)

Groups	Pathological scores	*P* value
Group A & asthma model group ($n = 5$)	6.25 ± 2.78	0.00[a]
Group A & control group ($n = 5$)	0.75 ± 0.96	
Group B & asthma model group ($n = 5$)	5.00 ± 0.41	0.00[a]
Group B & control group ($n = 4$)	0.40 ± 0.55	
Group C & asthma model group ($n = 4$)	4.33 ± 0.58	0.00[a]
Group C & control group ($n = 5$)	0.88 ± 1.03	
Group D & asthma model group ($n = 5$)	3.50 ± 0.41[b,c]	0.00[a]
Group D & control group ($n = 5$)	0.92 ± 0.83	

[a]indicated significant difference compared with control group

[b]indicated significant difference compared with group A

[c]*indicated significant difference compared with group B, P* < 0.05

Further, the serum IFN-γ levels and the IFN-γ/IL-4 ratio were higher in the group D asthmatic animals maintained in the general environment than the asthmatic animals of group A. this becomes significant given that asthma involves the role of Th2 cell-produced IL-4 and IL-5. Further, IFN-γ released by Th1 cells can shift the Th2 immune response towards Th1 to stall the development of asthma. This is overall suggestive of the vital involvement of intestinal microbes in altering IFN-γ levels or the IFN-γ/IL-4 ratio to quell the development of asthma (Qian et al, 2017).

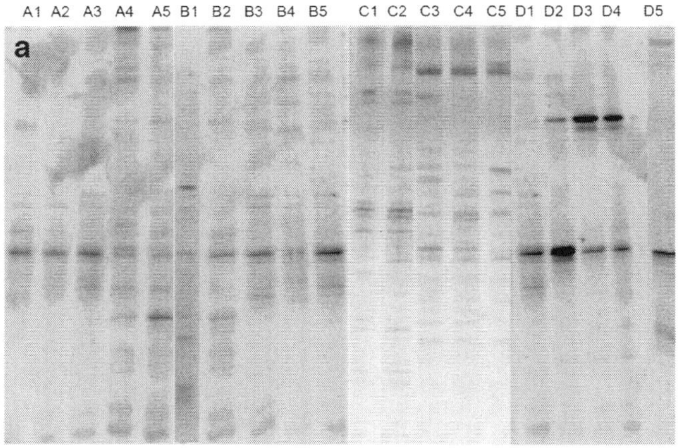

The DGGE profiles of intestinal flora of each group at three days after birth

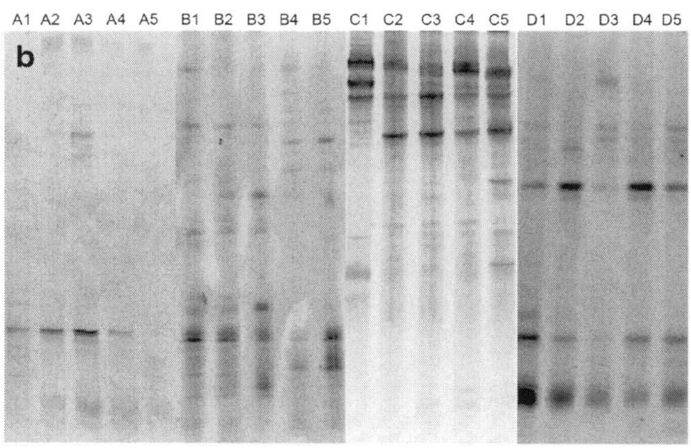

The DGGE profiles of intestinal flora of each group at three weeks after birth

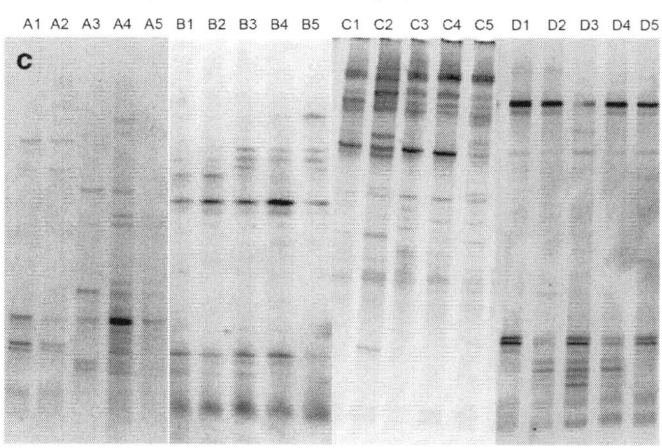

The DGGE profiles of intestinal flora of each group at five weeks after birth

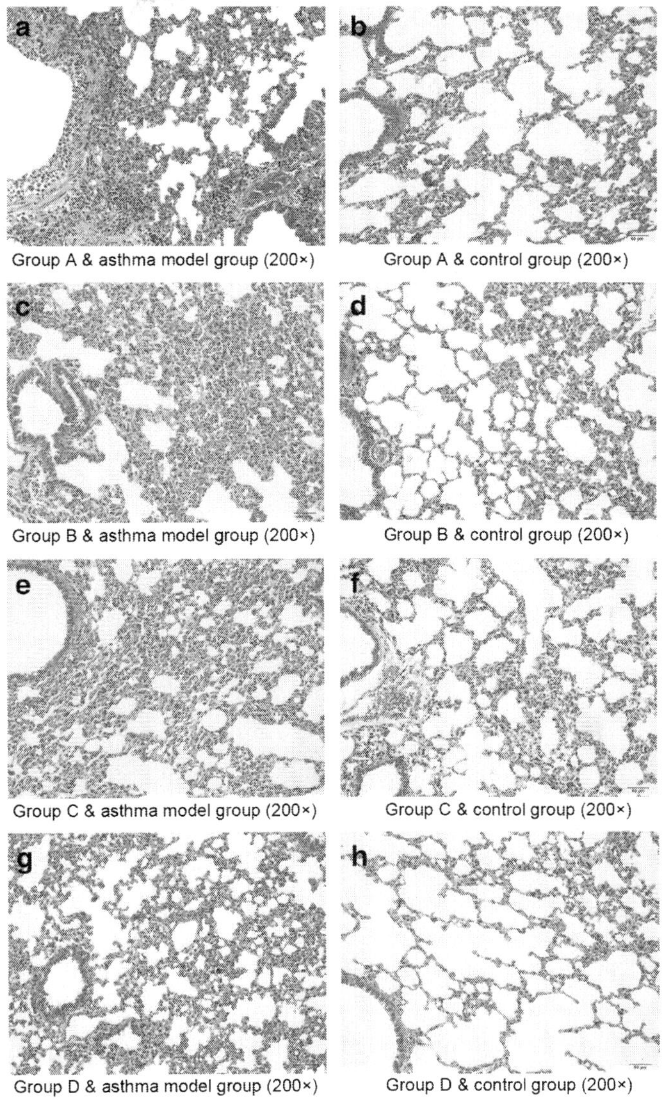

Figure 2: Above displays ***TOP:*** the diversity of intestinal microbes depending on early-age exposure. The greater is the abundance of the intestinal flora when the number of stripes is more. The species number is more with lighter stripes; ***BOTTOM:*** the pathological changes in mice following exposure to an allergen (OVA) (Qian et al, 2017: *Open-Access: This article is distributed under the terms of the Creative Commons Attribution 4.0 International License (http://creativecommons.org/licenses/by/4.0/), which permits unrestricted use, distribution, and reproduction in any medium, provided you give appropriate credit to the original author(s) and the source, provide a link to the Creative*

Commons license, and indicate if changes were made. The Creative Commons Public Domain Dedication waiver (http://creativecommons.org/publicdomain/zero/1.0/) applies to the data made available in this article, unless otherwise stated).

To quote the authors, "our findings indicate that exposure to diverse microbial environments in early life produces a clear impact on gut microbial colonization in BALB/c mice. Additionally, the diversity of the intestinal flora in early life may prevent airway inflammation in asthma via regulating the Th1/Th2 balance" (Qian et al, 2017).

It is interesting to note that between asthmatic and healthy people, the former display elevated interleukin (IL): IL-4, IL-5, and IL-13 levels. These cytokines are associated with Th2 type pathologies: atopic diseases that are tendencies to manifest allergies and high IgE levels upon ingestion or inhalation of an environmental antigen. These variations in addition to that of genetics and environmental factors can overall affect the gut microbiome (Frati et al, 2018).

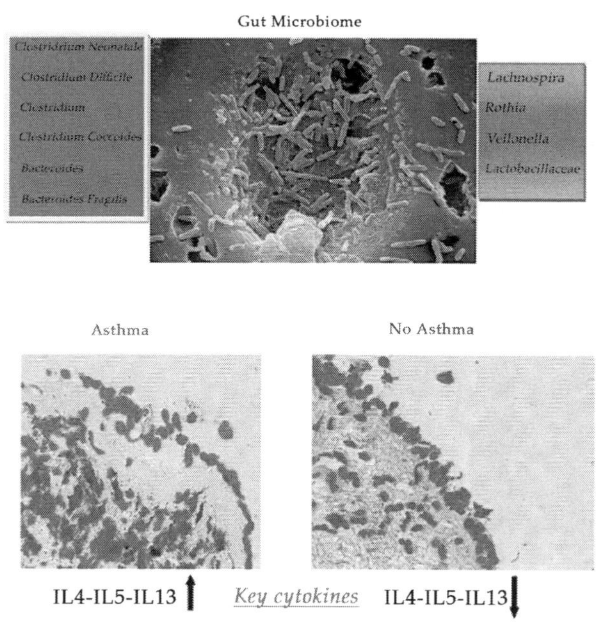

Figure 3: Above illustrates variations in the lung bacterial microbiota in children who are healthy and those with asthma. The pathogens involved in asthma development are shown on TOP while at the BOTTOM, the image on the left is the elevated profile of IL-4, IL-5, and IL-13 in a child with asthma and the right is a healthy child's respiratory tract (*Licensee MDPI, Basel, Switzerland; Frati*

The initial years of life strongly influence the gut microbiome to further play a possible role in immunity. In neonates, the gut colonization is impacted by several factors such as antibiotic exposure and cesarean section. Researchers Stokholm and team published their research in 2018 on the impact of gut colonization patterns and the risk of asthma using the Copenhagen Prospective Studies on Asthma in Childhood$_{2010}$ (COPSAC$_{2010}$) birth cohort. This cohort primarily has asthma as the primary outcome in detailed studies of 700 children.

Analyses of 16S rRNA in fecal samples revealed that in the first year of life, extensive gut microbiome changes occur. 2 microbial communities were identified by Partitioning around medoids (PAM) clustering into *Staphylococcus*, *Streptococcus*, *Bifidobacterium* and *Enterococcus* at ages of one-week and one-month and *Faecalibacterium*, *Bacteroides(x3)*, and *Anaerostipes* when the children were 1-year old.

While no evident relationships were unveiled between asthma risk and α-diversity (Shannon diversity and Chao1 indices), in terms of β-diversity, the microbial populations were different. Interestingly, there was no effect of the maternal asthma status that is suggestive of unwanted microbial effects on the risk of asthma. 20 genera were found to be abundant in the samples: *Bacteroides*, *Bifidobacterium*, and *Veillonella* being the most dominant before the incidence of asthma.

In the children borne by asthmatic mothers, the abundance of the genera *Faecalibacterium, Bifidobacterium, Roseburia, Alistipes, Lachnospiraceae incertae sedis*, *Ruminococcus*, and *Dialister* displayed a negative association to asthma at age 5 years. A positive correlation was noticed with *Veillonella*.

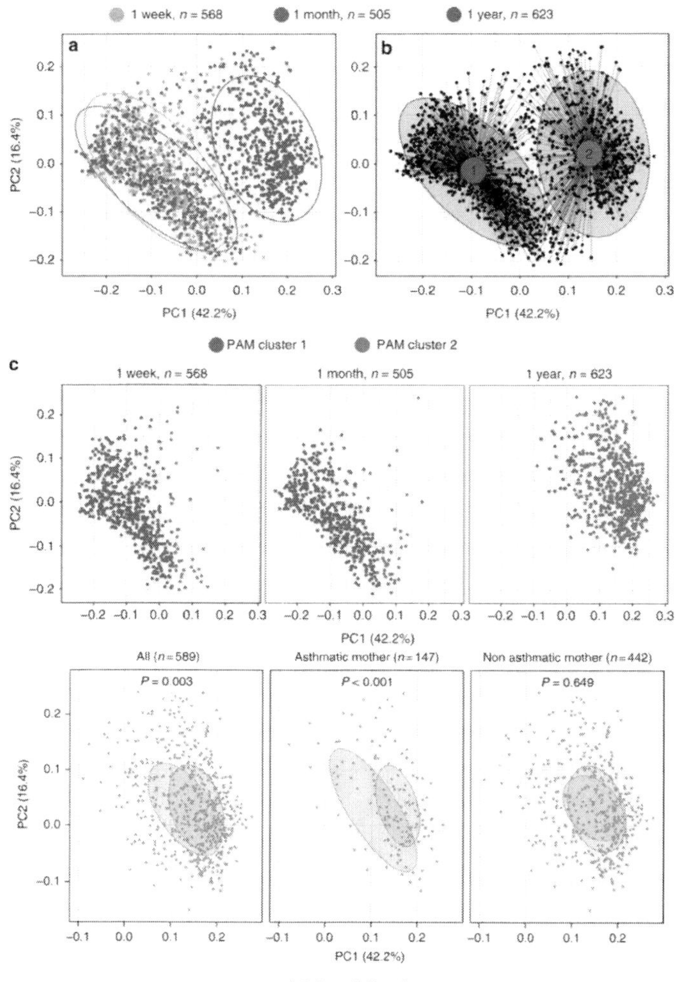

Figure 4: Above displays the impact of the gut microbiome development on the risk of asthma. ***TOP:*** Change in microbial composition in the initial year of life; ***BOTTOM:*** the association of later asthma and β-diversity in the 1-year fecal samples. The mean ± 1 SD of children, who at age 5 years is shown as ellipses: asthmatic (orange) (N=58) or non-asthmatic (green) (N=531). Subsets include asthmatic mother, asthmatic (orange) (N=25), and non-asthmatic (green) (N=122); non-asthmatic mother, asthmatic (orange) (N=33), and non-asthmatic (green) (N=409) (Stokholm et al, 2018: *Open-Access: This article is licensed under a Creative Commons Attribution 4.0 International License, which permits use, sharing, adaptation, distribution and reproduction in any medium or format, as long as you give appropriate credit to the original author(s) and the source, provide a link to the Creative Commons license, and indicate if changes*

Overall, the researchers concluded that as evidenced by microbial community type analyses, the risk of asthma involves the role of a skewed maturation of the gut bacteria in the first year of life (Stokholm et al, 2018).

Analyses based on next-generation sequencing (NGS) revealed that healthy people showed a predominance of Bacteroidetes, mainly the species Prevotella against those with asthma. Other bacterial phyla that are more associated with healthy lungs include Actinobacteria (species Corynebacterium), Firmicutes (species Streptococcus), and Proteobacteria (species Acinetobacter). An imbalance in the microbial diversity of dysbiosis is associated with the development of asthma (Hufnagl et al, 2020).

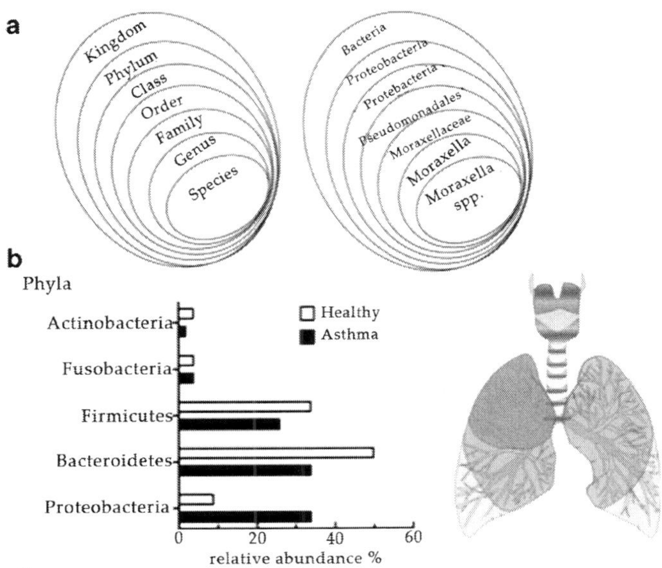

Genera

	Healthy	Asthma
Bacteroidetes	Prevotella	Prevotella↓
Firmicutes	Veillonella	Streptococcus↑
Proteobacteria	Acinetobacter	Moraxella ↑ Haemophilus↑
Actinobacteria	Corynebacterium	Corynebacterium ↓↓

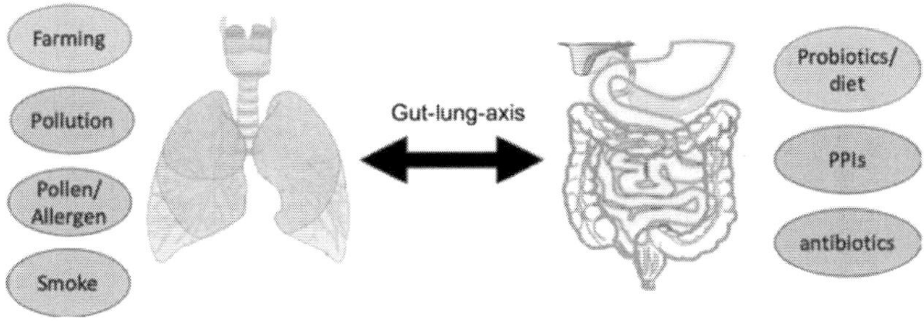

Figure 5: Above represents TOP: the ubiquitous phyla seen healthy and asthmatic individuals are shown as white and black bars. Asthmatic people show enrichment of Proteobacteria while Phyla Firmicutes, Actinobacteria, and Bacteroidetes and genera Prevotella and Corynebacterium are lower in asthmatics. BOTTOM: the gut-lung axis and the effect of various factors (Hufnagl et al, 2020: *Open Access: This article is licensed under a Creative Commons Attribution 4.0 International License, which permits use, sharing, adaptation, distribution and reproduction in any medium or format, as long as you give appropriate credit to the original author(s) and the source, provide a link to the Creative Commons licence, and indicate if changes were made. The images or other third party material in this article are included in the article's Creative Commons licence, unless indicated otherwise in a credit line to the material. If material is not included in the article's Creative Commons licence and your intended use is not permitted by statutory regulation or exceeds the permitted use, you will need to obtain permission directly from the copyright holder. To view a copy of this licence, visit http://creativecommons.org/licenses/by/4.0/*).

Treatment using proton pump inhibitors (PPIs) to target ulcers have shown to evidently cause dysbiosis. The abundance of *Lactobacillus* was lowered while that of *Streptococci* was enhanced in terms of the microbiota affecting the gut microbiota and increase the susceptibility to *Clostridium difficile* infections due to PPI use.

Table 2: below lists a comprehensive effect of the effects of drugs on the gut microbiome to cause dysbiosis or an imbalance in microbial diversity: *1 Alpha diversity: microbiota diversity in individual site or sample (one value per sample); 2 Diversity between separate samples; *PPI, proton pump inhibitor; **NSAID, non-steroidal anti-inflammatory drug* (Hufnagl et al, 2020: *Open Access: This article is licensed under a Creative Commons Attribution 4.0 International License, which permits use, sharing, adaptation, distribution and reproduction in any medium or format, as long as you give appropriate credit to the original author(s) and the source, provide a link to the Creative Commons licence, and indicate if changes were made. The images or other third party material in this article are included in the article's Creative Commons licence, unless indicated otherwise in a credit line to the material. If material is not included in the article's Creative Commons licence and your intended use is not permitted by statutory regulation or exceeds the permitted use, you will need to obtain permission directly from the copyright holder. To view a copy of this licence, visit http://creativecommons.org/licenses/by/4.0/*)

Category	**Drug**	**Strains affected**	**Alpha diversity[1]**	**Beta diversity[2]**	**Abundance**	**Organism**	**Notes**
Antibiotic	Vancomycin	*Firmicutes* ↓	↓		↓	Mouse, male, female	Colonic, 14 days after treatment
	Ciprofloxacin-metronidazole	*Firmicutes* ↓	↓		↓	Mouse, male	
		Bifidobacterium ↓	↓			Human	
	Vancomycin	*Firmicutes* ↑				Mouse, perinatal treatment	Hypersensitivity pneumonitis unaffected
	Streptomycin	*Bacteroidetes* ↑				Mouse	Hypersensitivity pneumonitis ↑
	Amoxicillin		n.s.			Human	5-day treatment course
	Azithromycin		↓			Human	
	Cotrimoxazole		n.s.			Human	
	Antibiotics mix	*Lachnospira* ↓	↓			Mouse	Intermittent exposure, 3 cycles. Reduction of pulmonary Tregs

PPI*		*Firmicutes* ↓/ *Erysipelotrichales or Clostridiales*	↓	Altered		Human	
		Actinomycetales ↑		Altered		Human	
		Micrococcaceae ↑		Altered		Human	
		Streptococcaceae ↑ /*Streptococcus pneumoniae* ↑		Altered	↑	Human	Gastric, intestinal, esophageal and oral microbiome; after short-term treatments
		While on PPI: *Lactobacillus* ↓ *Stenotrophomonas* ↓ *Haemophilus* ↑	On PPI: no change after PPI: ↑	On PPI: no change after PPI: ↑	After PPI: changed	Human, children	12 GERD infants treated 18 weeks with PPi. After discontinuation correlation to normal age/ nutrition
		Lactobacillus/L. gasseri subgroup, L. fermentum, the L. reuteri subgroup L. ruminis ↑					GERD patients, 4, 8 weeks treatment
		Faecalibacterium ↓	n.s.			Human	Long-term use over at least 1 year
		Clostridium difficile ↑					
		Bacteroidetes ↓ *Firmicutes* ↑ *Species: Holdemania filiformis* ↑, *Pseudoflavonifractor capillosus* ↓	n.s.	Moderate shift			Long-term use over more than 5 years
		Gammaproteobacteria ↑/*Enterobacter, Escherichia, Klebsiella, Citrobacter*		Altered		Human	
		Clostridium difficile infection. ↑			↑	Human	
		Bacterioidales/Bifidobacterium ↑			↑	Human	
		Enterococcaceae ↑			↑	Human	
		Gut commensals ↓, while oral and upper GI tract commensals ↑	↓			Human	Paired analysis: 70 monozygotic twin pairs
		Smithella	↓			Mouse	Dominant in responders (vs non-responders) to food allergen
		Acetivibrio	↓			Mouse	
		Faecalibacterium	↓			Mouse	
		Porphyromonadaceae / Tanerella, Barnesiella	↑			Mouse	

Other	Olanzapine		↓		Altered	Mouse	Acts bactericidic
	Metformin	*Bifidobacterium bifidum* ↑	No change		Altered	Human	
		E.coli ↑					
	NSAID**		No change		Not altered	Human	
	Opioids	*Staphylococcus* ↑ *Enterococcus* ↑	↑		Altered	Human	Due to obstipation

While the exact role played by gut microbes in asthma warrants in-depth studies, the vital involvement of the gut microbiota in regulating the immune system in asthma cannot be ruled out.

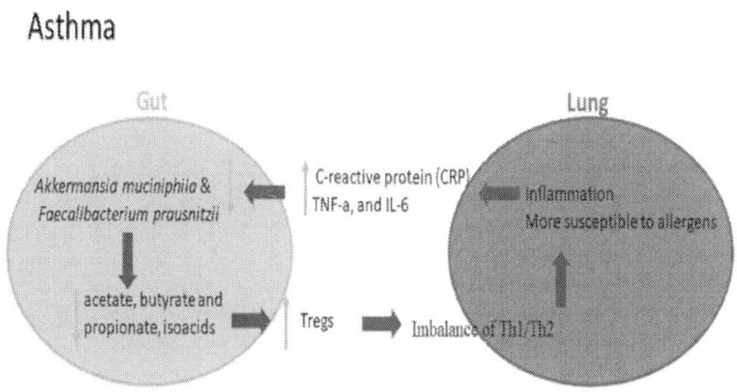

Figure 6: Above shows the involvement of the gut-lung axis in asthma (2020 Zhang, Li, Wang, Tan, Zhang and Feng: *This is an open-access article distributed under the terms of the Creative Commons Attribution License (CC BY). The use, distribution or reproduction in other forums is permitted, provided the original author(s) and the copyright owner(s) are credited and that the original publication in this journal is cited, in accordance with accepted academic practice. No use, distribution or reproduction is permitted which does not comply with these terms*).

2.3 COVID-19: THE PAUSE IN 2020

The word “2019 novel coronavirus disease (COVID-19)” induces chills in the minds of global citizens who have lived in the 21st century. The loss of lives and restrictions in the economy has left an indelible mark that may take longer to be erased.

This infectious disease is caused by severe acute respiratory syndrome coronavirus 2 (SARS-CoV-2). Wuhan in the Hubei province of China was the first site where the disease that has become a pandemic altering the health scenario of the world was first reported. Adults in the city known to be a major transportation hub of China reported to hospitals with severe pneumonia whose cause was not known. Several of these cases reported a common exposure to the Huanan wholesale seafood market known for also selling live animals.

The Wuhan Municipal Health Commission on 31st Dec 2019 reported to the World Health Organization or WHO that this was unknown pneumonia that was being subject to close monitoring. On 1st January 2020, the market was closed. The identification that it was a coronavirus was on 7th January. This virus showed a similarity of 70% to the SARS-CoV and around 95% homology with bat coronavirus.

The cases then rose up exponentially showing the ability of the virus to transmit among humans. 11th Jan 2020 was the day of recording the first human death. This was further fueled by the Chinese New Year associated migration. Cases were reported from people returning from Wuhan to Thailand, Japan, and South Korea. This was followed by reports of the disease being transmitted to healthcare workers on 20th Jan 2020. A "lockdown" was declared in Wuhan that was extended to other cities in the province.

On 30th January 2020, the WHO declared the outbreak to be a Public Health Emergency of International Concern.

Transmission among humans was established when people in other countries developed COVID-19 despite not traveling to China. This led to screening operations in airports to test for COVID-19 and isolating the symptomatic people. It was soon found that the virus could also be spread prior to symptom onset or from asymptomatic people.

11th March 2020 was the day the disease was declared as a pandemic with the number of people diagnosed touching 1 million on April 2, 2020. The fatality rate was 5.2% across 204 countries as against the SARS epidemic of 2002-2003 that infected 8096 people in 29 countries and the Middle East respiratory syndrome coronavirus (MERS-CoV) that targeted 2494 people in 27 countries between 2012-2019. As of July 1, 2020, more than 200 countries were affected by SARS-CoV-2 causing 10 million identified cases with 508 000 confirmed deaths (Chauhan, 2020; Singhal, 2020; Wiersinga et al, 2020; Shiv Sanjeevi Sripathi and Prerna Pandey, 2021).

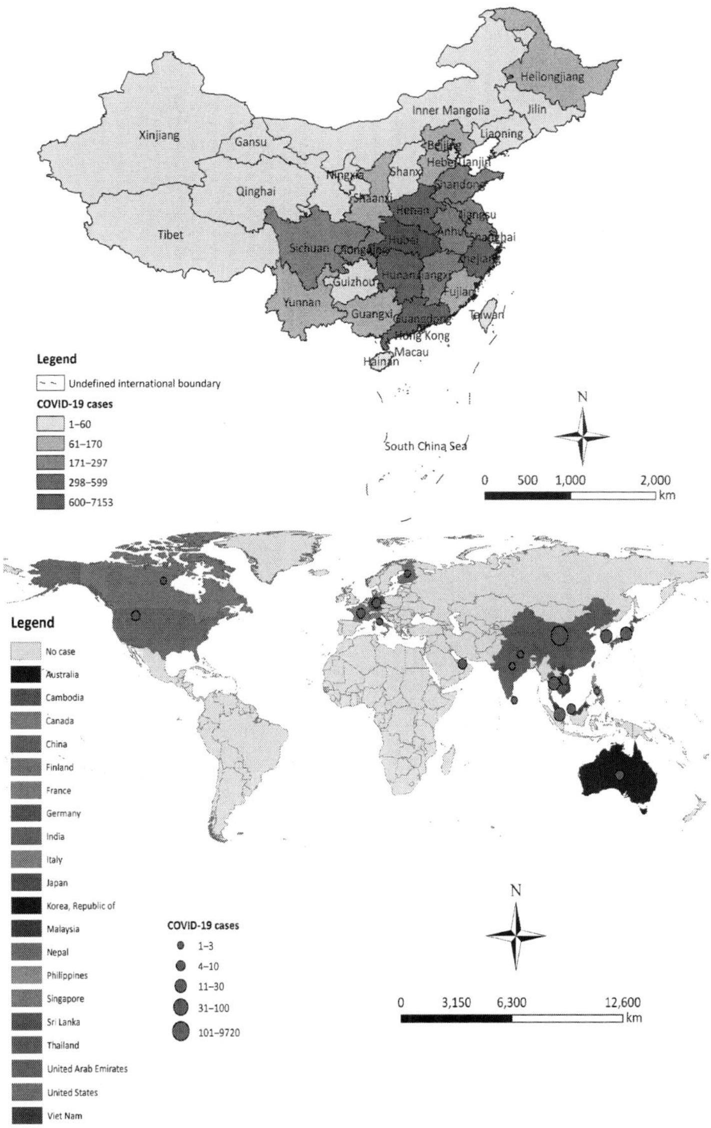

Figure 7: Above represents the spread of the virus that caused the global shutdown in 2020: ***TOP:*** Spread in China and ***BOTTOM:*** Global distribution (Adhikari **et al, 2020:** ***Open Access:*** *This article is licensed under a Creative Commons Attribution 4.0 International License, which permits use, sharing, adaptation, distribution and reproduction in any medium or format, as long as you give appropriate credit to the original author(s) and the source, provide a link to the Creative Commons licence, and indicate if changes were made. The images or other third party material in this article are included in the article's Creative Commons licence, unless indicated otherwise in a credit line*

Let us look at the major epidemiological indicators of COVID-19 research articles in January 2020 (Adhikari **et al, 2020:** ***Open Access:*** *This article is licensed under a Creative Commons Attribution 4.0 International License, which permits use, sharing, adaptation, distribution and reproduction in any medium or format, as long as you give appropriate credit to the original author(s) and the source, provide a link to the Creative Commons licence, and indicate if changes were made. The images or other third party material in this article are included in the article's Creative Commons licence, unless indicated otherwise in a credit line to the material: To view a copy of this licence, visit http://creativecommons.org/licenses/by/4.0/. The Creative Commons Public Domain Dedication waiver (http://creativecommons.org/publicdomain/zero/1.0/) applies to the data made available in this article, unless otherwise stated in a credit line to the data*):

Indicators	**Description**
Age of patients	• Cases range between 25 and 89 years, with most patients aged between 35 and 55 years and fewer cases among children and infants [14]
	• Median age of patients is 59 years, ranging from 51 to 89 [2]
	• Average age of patients was 55.5 years; age distribution: ≤ 39: 10%; 40–49: 22%, 50–59: 30%; 60–69: 22%, ≥ 70: 15% [19]
	• Cases range from 2 to 72 years [20]
Sex of patients	• More cases were males [20]
	• 59% males [2]
	• 68% males [19]
Age of the deaths	• Median age of death was 75 (with a range between 48 and 89 years) [21]
Exposure history	• Huanan Seafood Market in Wuhan [19, 22]
	• Wuhan residents or people who visited Wuhan [20]

Incubation time	• 4.8±2.6 days (2–11 days) [15]
	• 5.2 days (4.1–7 days) [2]
	• Average of 7 days (2–14 days) [23]
	• Average of 10 days [22]
	• 5–6 days [24]
	• Average of 6.4 days (5.6–7.7 days) [20]
Basic Reproduction (R_0)	• 2.6 (uncertainty range: 1.5–3.5) [25]
	• 3.8 (95% *CI*: 3.6–4.0) [26]

The pandemic disease displays a vast array of symptoms ranging from benign to fatal: some affected individuals display no symptoms while others can show infection of the upper airways to pneumonia and even acute respiratory distress syndrome (ARDS). Other complications are associated with acute kidney injury, shock, and cardiac issues. There have been reports of endothelial sloughing, pulmonary edema, and vascular thrombosis.

As of now, the support management administered to patients in the case of ADRS or acute hypoxic respiratory failure is advised. Supplemental oxygen therapy is required in 75% of the hospital admitted patients. If there is a lack of a response to this therapy, then high-flow nasal canula oxygen can be resorted to. In the case of several patients who receive mechanical ventilation, multi-organ failure is also seen that necessitates a long stay in the hospital (Zeppa et al, 2020; Shiv Sanjeevi Sripathi and Prerna Pandey, 2021).

Researchers Segal and team (2020) discussed how the gastrointestinal (GI) tract was impacted by COVID-19 severity. Points that support this include the presence of nausea, vomiting and diarrhea (GI symptoms) have also been reported as symptoms of the disease in addition to respiratory symptoms. Further, the shedding of the virus in the stool of patients and the detection of the pathogen in GI tract tissues further add weight to this observation.

Sequencing of the 16S ribosomal RNA (rRNA) gene V3-V4 region in a cross-sectional study of 30 COVID-19 patients, 24 influenza A (H1N1) patients, and 30 matched healthy controls by Gu and team (2020) revealed that the diversity and abundance of the gut microbiota were lower in the H1N1 and COVID-19 patients against the controls. The researchers reported that COVID-19 patients displayed an increase in taxa associated with higher C-reactive protein (CRP) values (the measurement of bacterial infection). These taxa were described as "opportunistic pathogens" inclusive of "*Streptococcus*, *Rothia*, *Veillonella*, *Erysipelatoclostridium*, and *Actinomyces*". *Streptococcus* and *Rothia* have been linked to secondary bacterial lung infection in patients with avian H7N9 virus infection while *Rothia* is possibly implicated in pneumonia seen in immunocompromised people and patients with retained catheters. The damage to worsen the manifestations of inflammatory bowel disease involves the role of *Actinomyces*. Further, H1N1 patients displayed a conspicuous lowering of the levels of butyrate-producing bacteria (BPB) that can cause an increase in intestinal pathogens to alter the intestinal cell permeability and cause the release of cytokines.

Another study by Zuo and team (2020) reported the analyses of fecal samples from 15 patients employing shotgun metagenomic sequencing vs. an equal number of controls. The severity of the disease described as mild, moderate, severe, or critical) was associated with 23 bacterial taxa of which 15 were the Firmicutes phylum. The numbers of beneficial commensals were low throughout hospitalization and remained so even after the virus was cleared and the respiratory symptoms were resolved. The viral load in fecal samples displayed an inverse association with *Bacteroides dorei*, *Bacteroides thetaiotaomicron*, *Bacteroides massiliensis*, and *Bacteroides ovatus. These bacteria are known to lower the expression of* angiotensin-converting enzyme 2 (ACE2) in the gut of mice. Interestingly, the angiotensin-converting enzyme 2 (ACE2) receptor present in the respiratory and GI tracts is the protein by which SARS-CoV-2 enters the host.

An amalgamation of responses is elicited as a response to the SARS-CoV-2 infection that includes B-cells, T-cells, macrophages, monocytes, and cytokine release. These factors have been profiled by research and reported to be tumor necrosis factor (TNF), interleukin-2, interleukin-6, interleukin-7, and interleukin-10 (IL-2, IL-6, IL-7, and IL-10) along with monocyte chemoattractant protein-1 (MCP1), granulocyte colony–stimulating factor (G-CSF), macrophage inflammatory protein 1α (MIP1α), and interferon-γ-inducible protein-10 (IP-10) (Zeppa et al, 2020).

Given the in-depth insights offered by the OMICS approach, Guo and team (2020) reported that the application of a machine learning model could allow core micribiota of the gut to predict proteomic biomarkers that were found predict the progression of COVID-19. The main operational taxonomic units (OTUs) were identified as "*Bacteroides* genus, *Streptococcus* genus, *Lactobacillus* genus, *Ruminococcaceae* 120 family, *Lachnospiraceae* family and *Clostridiales* order". The gut bacteria OTUs were found to outperform other factors such as demographic and lab parameters in predicting the

proteomic -risk score (PRS) of worsening of the disease, as shown by correlation analysis.

The dietary fiber and resistant starch in food can be processed by gut microbiota to form acetate, propionate, and butyrate (short-chain fatty acids or SFCAs). These metabolites have been implicated in a broad array of functions ranging from an energy source for intestinal cells to targeting LPS-mediated inflammation, cell activation, and chemotaxis (Zeppa et al, 2020).

This becomes relevant with few studies, for example, that reported by Trompette and team (2014) who showed that the ratio of Firmicutes to Bacteroidetes in the gut was altered by the content of dietary fermentable fiber. The fiber was metabolized into SFCAs and mice that were administered a diet rich in fiber showed higher levels of SFCAs in their blood. This increased SFCA profile exerted a protective effect from allergic inflammation as opposed to animals that were on a low-fiber diet.

This brings in the possible vital involvement of SFCAs in attenuating inflammation by targeting immune cell proliferation and cytokine release.

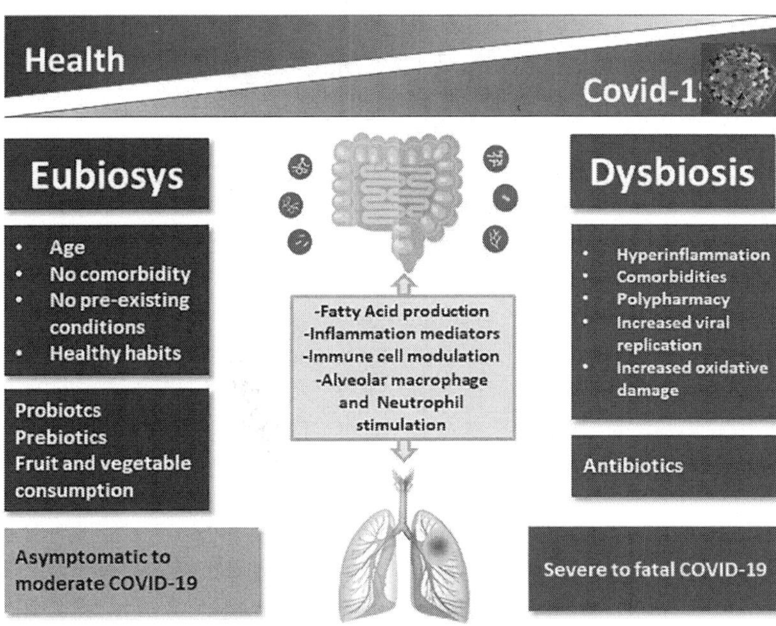

Figure 8: Above is a detailed and well-presented scheme of how the microbiota changes in COVID-19 patients (2020 Donati Zeppa, Agostini, Piccoli, Stocchi and Sestili: *This is an open-access article distributed under the terms of the*

Further studies that are looking at the association of the gut microbiota in COVID-19 are summarized below in Table 3: *COVID-19, Coronavirus disease; ICU, intensive care unit; Interv., interventional; Obs., observational; Par. Ass., parallel assignment; Prosp., prospective; GM, gut microbiota.* † Except those indicated, all the trials are Phase "Not Applicable", which is a definition used to describe trials without Food and Drug Administration (FDA)-defined phases. * Estimated total number of participants (Ferreira et al, 2020: Licensee MDPI, Basel, Switzerland. *This article is an open-access article distributed under the terms and conditions of the Creative Commons Attribution (CC BY) license (http://creativecommons.org/licenses/by/4.0/: You are free to Share — copy and redistribute the material in any medium or format; Adapt — remix, transform, and build upon the material for any purpose, even commercially. The licensor cannot revoke these freedoms as long as you follow the license terms*):

ID, Country, and Status †	**Study Type and Participants**	**Trial Title (Main Hypothesis/Aims)**
NCT04325919 Hong Kong Recruiting	Obs./Prosp. Hospitalized COVID-19 patients (170 *)	Comprehensive Clinical, Virological, Microbiological, Immunological and Laboratory Monitoring of Patients Hospitalized With COVID-19
NCT04359706 France Recruiting	Obs./Prosp. COVID-19 patients at ICU (30 *)	Bacterial and Fungal Microbiota of Patients with Severe Viral Pneumonia With SARS-CoV2 (determine the respiratory and fecal microbiota—microbial and fungal—of critically ill patients)
NCT04410263 Switzerland Recruiting	Obs./Prosp. COVID-positive patients at ICU (300 *)	Microbiota in COVID-19 Patients for Future Therapeutic and Preventive Approaches
NCT04486482 USA Recruiting	Interv./Par. Ass. Mild-to-moderate COVID-19 patients (50 *)	An Exploratory, Open Label, Clinical Study to Evaluate the Physiologic Effects of KB109 in Adult Patients with Mild-to-Moderate COVID-19 on Gut Microbiota Structure and Function in the Outpatient Setting
NCT04399252 USA Recruiting	Interv./Par. Ass. Exposed household contacts of COVID-19 (1000 *)	A Randomized Trial of the Effect of Lactobacillus on the Microbiome of Household Contacts Exposed to COVID-19
NCT04355741 Portugal Recruiting	Obs./Prosp. COVID-19 patients (60 *)	Gut Microbiota, "Spark and Flame" of COVID-19 Disease (fecal gut microbiota composition could affect vulnerability and disease outcomes of COVID-19)

NCT04366089 Italy Recruiting	Interv./Par. Ass. Hospitalized COVID-19 patients (152 *)	Oxygen-Ozone as Adjuvant Treatment in Early Control of Disease Progression in Patients With COVID-19 Associated with Modulation of the Gut Microbial Flora (Phase 2)
NCT04332016 France Recruiting	Obs./Prosp. Hospitalized COVID-19 patients (2000 *)	COVID-19 Biological Samples Collection (COL-COV19-BX)
NCT04458519 Canada Not yet Recruit-ing	Interv./Par. Ass. COVID-19 patients not requiring hospitaliz. (40 *)	Randomised Single Blinded Clinical Study of Efficacy of Intranasal Probiotic Treatment to Reduce Severity of Symptoms in COVID19 Infection
NCT04327570 Belgium Recruiting	Obs./Prosp. Hospitalized COVID-19 patients (100 *)	In-depth Characterisation of the Dynamic Host Immune Response to Coronavirus SARS-CoV-2 (Correlation of immune profiling with microbiome analysis)
NCT04390477 Spain Recruiting	Interv./Par. Ass. Hospitalized COVID-19 patients (40 *)	The Intestinal Microbiota as a Therapeutic Target in Hospitalized Patients With COVID-19 Infection (a positive effect of probiotic on the GM that could produce a less severe clinical evolution of the disease)
NCT04368351 Italy Active, not recruiting	Obs./Retros. Hospitalized COVID-19 patients (70 *)	Evaluation of the Impact of Bacteriotherapy in the Treatment of COVID-19 (probiotic supplementation (SivoMixx) + Azithromycin)
NCT04373148 USA Recruiting	Obs./Prosp. Adults and children with COVID-19 (1000 *)	Understanding Immunity to SARS-CoV-2, the Coronavirus Causing COVID-19
NCT04451577 Italy Recruiting	Obs./Case-Control Hospital employees with or without COVID-19 (5000 *)	Epidemiologic, Clinical, Molecular Characteristics of Hospital Employees With or Without Covid-19 Infection: a Retrospective-prospective Cohort Study
NCT04497402 Italy Not yet Recruit-ing	Obs./Prosp. Covid-19 patients (88 *)	Sex-Informed Data in the COVID-19 Pandemic (determine whether there are sex differences in biomarkers, including in gut microbiome)
NCT04359459 The Netherlands Not yet recruiting	Obs./Prosp. COVID-19 patients (150 *)	Nasal CIliated EPithelial Genetic and Single Cell RNA prOfiLes of miLd, Severe and Very Severe COVID-Nineteen patIents (CIPOLLINI) Study (correlation of feces microbiome and clinical outcome for COVID-19)
NCT04403646 Argentina Not Yet Recruit-ing	Interv./Par. Ass. Hospitalized COVID-19 patients (140 *)	Efficacy of Tannin Specific Natural Extract for Coronavirus Disease (COVID-19): Randomized Controlled Trial
NCT04517422 Mexico Not Yet Recruit-ing	Interv./Par. Ass COVID-19 patients with mild symptoms (300 *)	Efficacy and Safety of Lactobacillus Plantarum and P. Acidilactici as Co-adjuvant Therapy for Reducing the Risk of Severe Disease in Adults With SARS-CoV-2 and Its Modulation of the Fecal Microbiota: A RCT
NCT04447144 Egypt Recruiting	Obs./Prosp.; COVID-19 patients with mild and moderate severity (200 *)	Nutritional Habits, Does it Affect Coronavirus Disease 2019 (COVID-19) Infection Outcome? An Egyptian Experience

NCT04420676 Austria Not yet recruiting	Interv./Par. Ass. COVID-19 patients (108 *)	Synbiotic Therapy of Gastrointestinal Symptoms During Covid-19 Infection: A Randomized, Double-blind, Placebo Controlled, Telemedicine Study (SynCov Study) (probiotic supplementation: Omnibiotic® AAD)
NCT04345510 Germany Not yet recruiting	Obs./Prosp. Asymptomatic Carriers (500 *)	Testing for COVID-19 Infection in Asymptomatic Persons (analyze for a possible correlation between oral microbiome and COVID-19 infection status)
NCT04444609 United Kingdom Recruiting	Obs./Prosp. COVID-19 patients with lung disease (230 *)	PROSAIC-19 - Prospective Longitudinal Assessment in a COVID-19 Infected Cohort
NCT04359836 USA Recruiting	Obs. COVID-19 patients (250 *)	A Non-Interventional Pilot Study to Explore the Role of Gut Flora in COVID-19 Infection

SARS CoV targets not only the lungs but also as mentioned earlier the GI tract. The disease is associated with a triad, according to Ferreira et al (2020) involving inflammation, hyper response, and the gut microbiota. This and the aforementioned few studies open up the possible involvement of the gut microbiota to intricately impact the susceptibility to COVID-19 via immune responses along with the response to treatments administered.

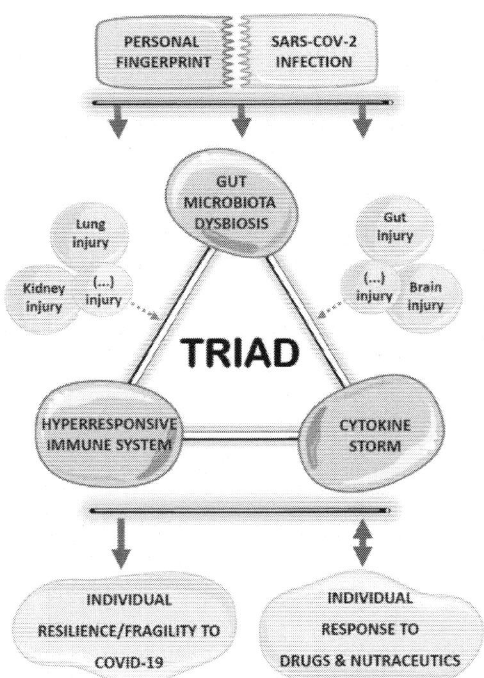

Figure 9: Above is a scheme of the triad involving gut bacteria that intricately control the pandemic of 2019-20 (Ferreira et al, 2020: Licensee MDPI, Basel,

An exciting development is the possibility of modulating the gut bacteria to target the inflammation seen in COVID-19. According to McIlroy and team (2020), the possibility of fecal microbiota transplantation or FMT or **intestinal microbiome transfer** can be explored following intensive care and evaluation. As an example, a previously published article by Wei and team (2016) in using this approach to address the symptoms of 2 patients who developed multiple organ dysfunction syndrome (MODS). This complex condition is possibly caused by trauma, shock, and inflammation with the involvement of microbes from the gastrointestinal tract. 16S rDNA-based molecular analysis of fecal microbiota revealed that the microbes were dysregulated in a 65-year-old with cerebellar hemorrhage and an 84-year-old with cerebral infarction. FMT entailed one nasogastric infusion of sterile-filtered, pathogen-free feces from a healthy donor followed by relevant monitoring. Both patients displayed a significant lowering of MODS and severe diarrhea with normalization of the body temperature and stool output. Additionally, the count of *Proteobacteria* (a phylum of opportunistic organisms) was lowered while that of *Firmicutes* (intestinal commensals) was increased along with a decrease in inflammatory factors evaluated viz. IL-6, CRP, PCT, and ESR. Overall, the complications by correcting microbial dysregulation associated with sepsis were lowered by this approach using FMT.

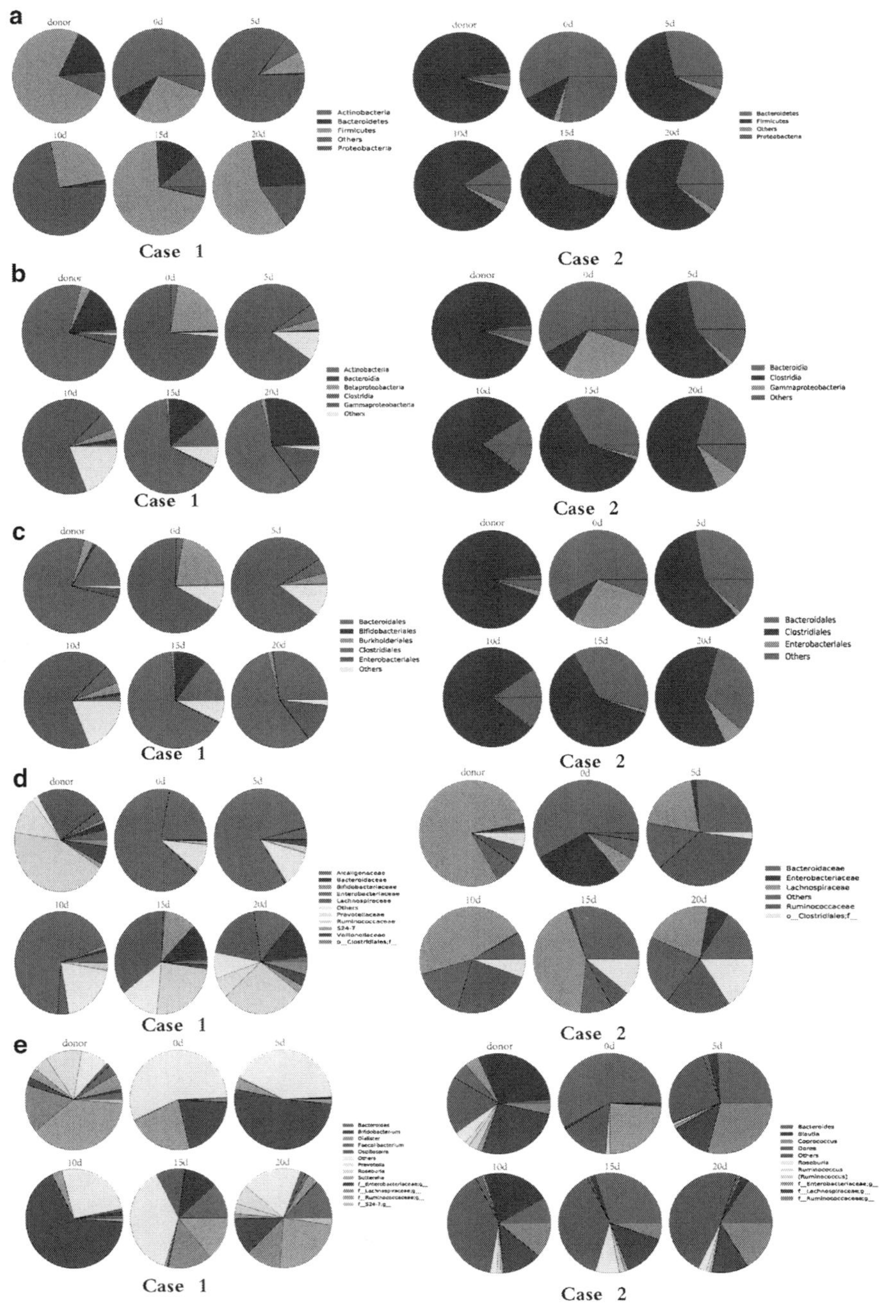
a
donor
0d
5d
10d
15d
20d
Actinobacteria
Bacteroidetes
Firmicutes
Others
Proteobacteria
Case 1
Bacteroidetes
Firmicutes
Others
Proteobacteria
Case 2
b
Actinobacteria
Bacteroidia
Betaproteobacteria
Clostridia
Gammaproteobacteria
Others
Case 1
Bacteroidia
Clostridia
Gammaproteobacteria
Others
Case 2
c
Bacteroidales
Bifidobacteriales
Burkholderiales
Clostridiales
Enterobacteriales
Others
Case 1
Bacteroidales
Clostridiales
Enterobacteriales
Others
Case 2
d
Alcaligenaceae
Bacteroidaceae
Bifidobacteriaceae
Enterobacteriaceae
Lachnospiraceae
Others
Prevotellaceae
Ruminococcaceae
S24-7
Veillonellaceae
o__Clostridiales;f__
Case 1
Bacteroidaceae
Enterobacteriaceae
Lachnospiraceae
Others
Ruminococcaceae
o__Clostridiales;f__
Case 2
e
Bacteroides
Bifidobacterium
Dialister
Faecalibacterium
Oscillospira
Others
Prevotella
Roseburia
Sutterella
f__Enterobacteriaceae;g__
f__Lachnospiraceae;g__
f__Ruminococcaceae;g__
f__S24-7;g__
Case 1
Bacteroides
Blautia
Coprococcus
Dorea
Others
Roseburia
Ruminococcus
[Ruminococcus]
f__Enterobacteriaceae;g__
f__Lachnospiraceae;g__
f__Ruminococcaceae;g__
Case 2

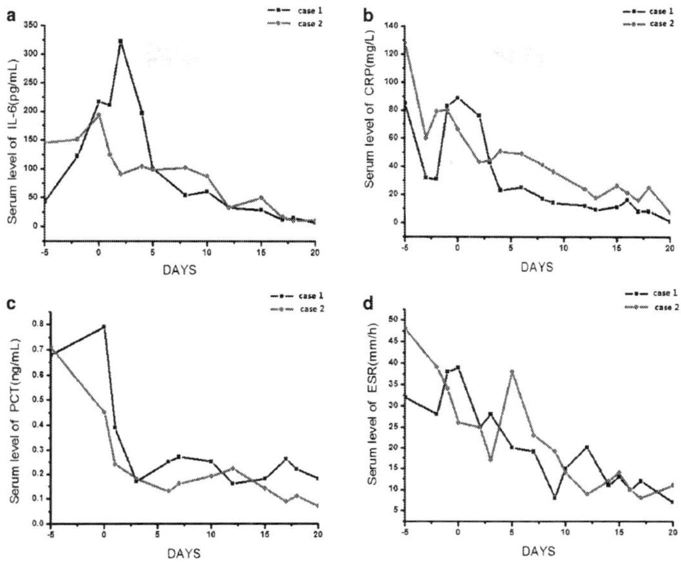

Figure 10: below shows TOP: the microbiota composition in the stool of the healthy donor vs. that of the patients who received FMT at the level of (a) phyla (b) the class (c) the order level (d) the family, and (e) the genus level. The days following the transplant of fecal microbiota are shown in days (*d*); BOTTOM: The lowering of inflammatory markers following the transplant of fecal microbiota (Wei et al, 2016: ***Open-Access*:** *This article is distributed under the terms of the Creative Commons Attribution 4.0 International License (http://creativecommons.org/licenses/by/4.0/), which permits unrestricted use, distribution, and reproduction in any medium, provided you give appropriate credit to the original author(s) and the source, provide a link to the Creative Commons license, and indicate if changes were made. The Creative Commons Public Domain Dedication waiver (http://creativecommons.org/publicdomain/zero/1.0/) applies to the data made available in this article, unless otherwise stated*).

Researchers McIlroy et al (2020) state, "When taken collectively, the available data insinuates that FMT - and indeed other less complex microbiome modulating therapeutic modalities, such as live biotherapeutic products (LBPs) - could be useful adjuvant immunomodulatory interventions when administered alongside agents currently under investigation, such as tocilizumab or dexamethasone, in patients with COVID-19 induced hyperinflammation".

2.4 CYSTIC FIBROSIS

Caucasians record cystic fibrosis (CF) in 1 in 3,000 – 4,000 live births while this figure is lower for Latin Americans (1 in 4,000 – 10,000) and African Americans (1 in 15,000 – 20,000) (Sanders and Fink, 2016). The disease involves the formation of thick mucus due to the drying of the airway surface liquid layer that in turn is due to an aberrant flow of chloride and bicarbonate ions across the epithelial cells. The mutation of the cystic fibrosis transmembrane conductance regulator (CFTR) gene results in this condition. While the most severely affected organ is the lung, the presence of the CFTR protein in the GI tract also affects the GI system (Davis, 2006).

A 2013-published article described the "first link" between fecal microbiota alterations and CF. One of the early sites where CF manifests is the GI tract, with an affected quality of life due to an excessive load of bacteria in the small intestine referred to as small bowel bacterial overgrowth (SBBO). The composition of the fecal microbes was analyzed employing Temporal Temperature Gradient Electrophoresis and species-specific PCR in CF patients. Genotyping of 36 patients for the ubiquitous F508del mutation followed by assigning them as no F508del mutation (n=12), homozygous F508del (n=5), and heterozygous F508del (n=19).

It was observed that the fecal microbiota composition was influenced by the CFTR mutations: whether they lead to a mild or severe clinical phenotype. The evenness (*Esi*) and biodiversity (*Hsi*) of the fecal microbiota were more in homozygous F508del mutation samples against the ones who had no F508del and to heterozygous-F508del. The homozygous-F508del and severe CF patients displayed a lower count of *Faecalibacterium prausnitzii*, *Bifidobacterium spp.*, and *Eubacterium limosum (beneficial bacteria) while Escherichia coli* and *Eubacterium biforme (probably harmful). Escherichia coli* plays a role in IBD and Celiac disease while there are reports of *E. biforme being isolated from wounds.*

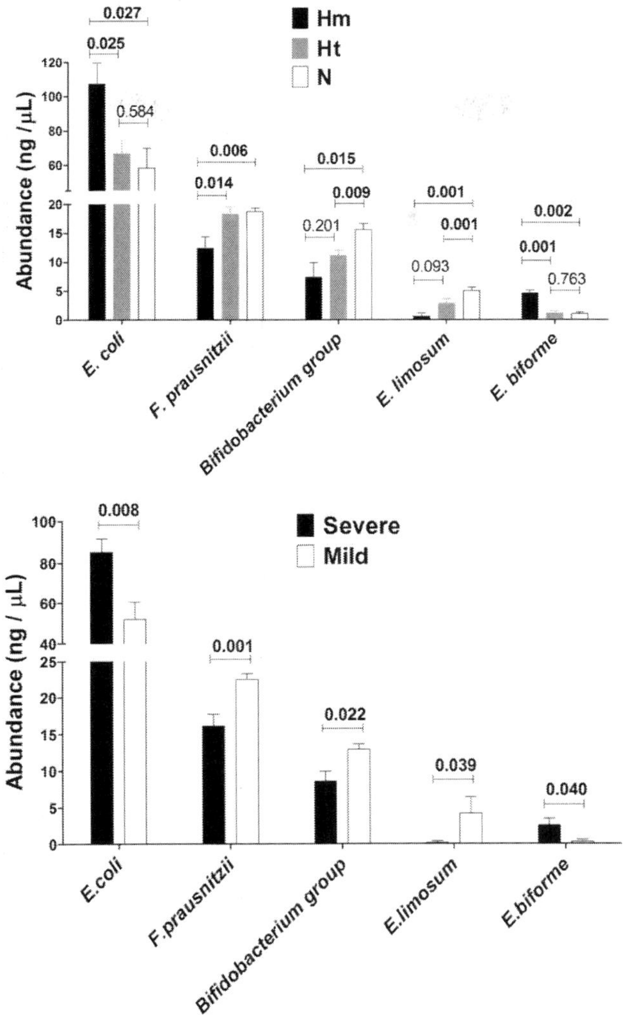

Figure 11: below shows the results of the species analyses with ***TOP***: homozygous-F508del patients showing higher levels of *E. coli* and *E. biforme while* non-F508del patients displayed more levels of *F. prausnitzii*, *Bifidobacterium*, and *E. limosum: black bars:* homozygous-F508del (Hm), grey bars represent heterozygous-F508del (Ht) and white bars represent non-F508del (N); *BOTTOM: The elevated profile of F. prausnitzii*, *Bifidobacterium*, and *E. limosum in patients with mild CF (white bars) vs. higher numbers of E. coli* and *E. biforme in patients with serious disease (black bars) (*Schippa et al, 2013:*This is an open-access article distributed under the terms of the Creative Commons Attribution License, which permits unrestricted use, distribution, and reproduction in any medium, provided the original author and source are properly credited*).

Researchers Fouhy and team (2017) conducted a pilot study to look at alterations in gut microbiota functionality in adult CF patients. The study entailed the extraction of metagenomic DNA from six patients with CF and six controls that were subjected to metagenomic sequencing using the MiSeq platform while metabolomics involved the use of gas chromatography-mass spectrometry.

Analyses using the HUMAnN2 analysis pipeline revealed that the CF group displayed significant abundances in 10.7% of 2066 pathways. CF patients displayed an increase in lipid metabolism pathways: biosynthesis of saturated/unsaturated fatty acids, fatty acid elongation, and β-oxidation. This was attributed to higher levels of *Ruminococcus*, *Clostridia*, *Enterococcus* and *Eggerthella* in CF patients showing that slight alterations of the microbiota can make a big impact on the functionality. *Ruminococcus gnavus*, *R. torques*, *C. clostridioform* and *E. faecalis* accounted for the higher usage of the lactose utilization pathway in the gut of CF patients. The possible functioning of *Bacteroides, Lachnospiracheae, Clostridium and Enterococcus* whose abundance was more in patients could account for an increase in the starch degradation pathway in the gut. Xenobiotic metabolism pathways (given the administration of drugs in CF patients vs. controls), folate metabolism, and the pentose phosphate pathway were increased in the gut of CF patients.

Further, it was revealed that *Clostridium*, *Blautia*, *Lachnospiraceae*, *Bacteroides* and *Enterococcus* in the gut of patients displayed an abundance of drug transmembrane transporter activity, penicillin binding, porin activity, and O acetyltransferase activity.

Both the patients and controls had comparable levels of viruses and *Verrucomicrobia*. The CF group did not show *Archaea* while the levels of *Actinobacteria*, *Bacteroidetes* and *Proteobacteria* were decreased and that of *Firmicutes* was increased compared to the non-CF controls. Further, only the CF group showed the presence of *Enterococcus faecalis* and *Clostridium hathewayi*: the former is known for its antibiotic resistance and the latter is linked to infections.

Table 4: below shows the alterations in the gut microbiota of CF patients vs. non-CF controls (Fouhy et al, 2017: *Open-Access: This article is licensed under a Creative Commons Attribution 4.0 International License, which permits use, sharing, adaptation, distribution and reproduction in any medium or format, as long as you give appropriate credit to the original author(s) and the source, provide a link to the Creative Commons license, and indicate if changes were*

Phylogeny	**CF average relative abundance**	**Control average relative abundance**	**P value (corrected)**
Actinobacteria			
Bifidobacterium	3.20967333	6.81271667	0.05
Bifidobacterium longum	2.12066833	4.23268333	0.046
*Bifidobacterium longum*_unclassified	2.12066833	4.23268333	0.046
Eggerthella lenta	0.19998833	0.00330333	0.013
Eggerthella lenta|t__ GCF_000024265	0.19998833	0.00330333	0.013
Bacteroidetes			
Barnesiella	0	1.514735	0.007
Barnesiella intestinihominis	0	1.514735	0.007
Barnesiella_intestinihominis|t__ GCF_000296465	0	1.514735	0.007
Alistipes putredinis	0.05026667	4.24946833	0.05
Alistipes putredinis|t__ GCF_000154465	0.05026667	4.24946833	0.05
Firmicutes			
Bacilli	12.81672	0.78161167	0.016
Lactobacillales	12.456185	0.78161167	0.016
Enterococcaceae	9.54589833	0	0.007
Enterococcus	9.54589833	0	0.007
Enterococcus faecalis	8.832115	0	0.007
*Enterococcus faecalis*_unclassified	8.832115	0	0.007
Lactobacillaceae	1.59546167	0	0.022
Lactobacillus	0.483995	0	0.022
Streptococcus thermophilus	1.27870833	0.10655	0.049
*Streptococcus thermophilus*_unclassified	1.27870833	0.10655	0.049
Clostridiaceae	11.9238667	0.50407667	0.004
Clostridium	11.9238667	0.48078667	0.004
Clostridium clostridioforme	6.65942167	0.11350667	0.022

*Clostridium clostridioforme*_unclassified	6.65942167	0.11350667	0.022
Clostridium hathewayi	0.29948667	0	0.007
*Clostridium hathewayi*_unclassified	0.29948667	0	0.007
Clostridiales _noname	0.15274667	0.00104167	0.013
Flavonifractor	0.14014	0.00104167	0.05
Flavonifractor plautii	0.14014	0.00104167	0.05
Flavonifractor plautii GCF_000239295	0.14014	0.00104167	0.05
*Anaerostipes*_unclassified	0.28471833	0	0.007
Blautia	10.2605633	0.89512667	0.016
Blautia producta	0.77242	0	0.022
Blautia producta|t__ GCF_000373885	0.77242	0	0.022
Ruminococcus gnavus	8.05991833	0.38045667	0.022
*Ruminococcus gnavus*_unclassified	8.05991833	0.38045667	0.022
Ruminococcus torques	1.07679	0.054245	0.013
*Ruminococcus torques*_unclassified	1.07679	0.054245	0.013
Coprococcus	0.17310667	4.51670167	0.033
Coprococcus comes	0.13519667	2.253635	0.021
Coprococcus comes|t__ GCF_000155875	0.13519667	2.253635	0.021
Dorea formicigenerans	0.169365	0.73247667	0.049
*Dorea formicigenerans*_unclassified	0.169365	0.73247667	0.049
*Lachnospiraceae bacterium*_1_1_57FAA	0.40623667	0	0.007
*Lachnospiraceae bacterium*_1_1_57FAA|t__ GCF_000218445	0.40623667	0	0.007
*Lachnospiraceae bacterium*_5_1_63FAA	0.01507333	0.414975	0.05
*Lachnospiraceae bacterium*_5_1_63FAA|t__ GCF_000185525	0.01507333	0.414975	0.05
*Lachnospiraceae bacterium*_8_1_57FAA	0.15276667	0	0.022
*Lachnospiraceae bacterium*_8_1_57FAA|t__ GCF_000185545	0.15276667	0	0.022
Faecalibacterium	0.06931833	3.33589167	0.003
Faecalibacterium prausnitzii	0.06931833	3.33589167	0.003
*Faecalibacterium prausnitzii*_unclassified	0.06931833	3.33589167	0.003

The increase in fat metabolism in CF patients can be attributed to a diet rich in fat administered to compensate the poor fat absorption. The high administration of antibiotics to CF patients resulted in the genes related to antibiotic processing and porin activity being enriched to facilitate the adaptation of the gut bacteria to the drug load.

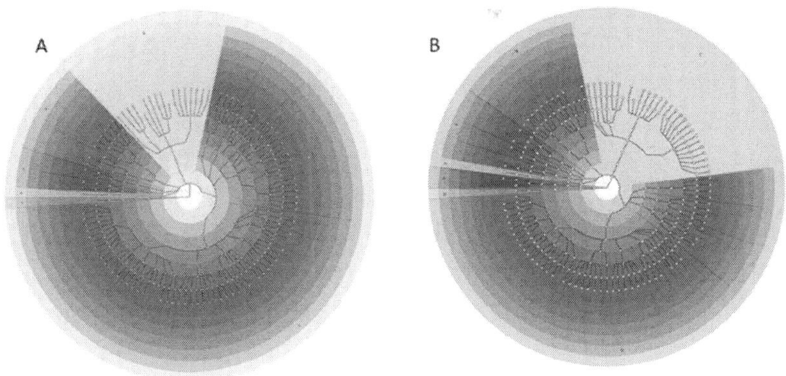

Figure 12: Above shows the phylogenetic composition as generated by 16S rRNA analyses (A) CF patients (B) controls: A: *Actinobacteria* (red), B: *Bacteroidetes* (yellow), C: *Firmicutes* (green), D: *Proteobacteria* (purple), E: *Verrucomicrobia* (orange), F: *Viruses* (grey). Control legend: A: *Archaea* (blue), B: *Actinobacteria* (red), C: *Bacteroidetes* (yellow), D: *Firmicutes* (green), E: *Proteobacteria* (purple), F: *Verrucomicrobia* (orange), G: *Viruses* (grey) (Fouhy et al, 2017: *Open-Access: This article is licensed under a Creative Commons Attribution 4.0 International License, which permits use, sharing, adaptation, distribution and reproduction in any medium or format, as long as you give appropriate credit to the original author(s) and the source, provide a link to the Creative Commons license, and indicate if changes were made. The images or other third party material in this article are included in the article's Creative Commons license, unless indicated otherwise in a credit line to the material. If material is not included in the article's Creative Commons license and your intended use is not permitted by statutory regulation or exceeds the permitted use, you will need to obtain permission directly from the copyright holder. To view a copy of this license, visit http://creativecommons.org/licenses/by/4.0/*).

2.5 CHRONIC OBSTRUCTIVE PULMONARY DISEASE (COPD)

Moderate to severe COPD is reported in 65 million people as per the World Health Organization with 3 million people having lost of lives due to this condition in 2005. Estimates are pitched that COPD will rank as the third leading cause of death by 2030 (Koul, 2013). The disease is attributed to majorly exposure to pollutants in the air (for example, smoking) along with genetics and immune pathways. The manifestations such as emphysema and chronic bronchitis adversely impact the functioning of the lungs. Research suggests the vital role of elevated levels of molecules of the immune system such as IL-6 and CRP. The former is produced as a response to smoke or bacteria while the latter is induced by IL-6 as a late-phase protein (Rabe and Watz, 2017; Zhang et al, 2020).

There are indications of the liver in this disease on account of the crucial involvement of this organ in amplifying the aforementioned immune responses. Studies have also pointed out how a diet rich in fiber lowered the risk of COPD. This brings in the role of the gut microbes given that most of them utilize fiber in the diet. This gut-liver-lung axis becomes vital to synthesize SFCAs and maintain the intestinal barrier functions.

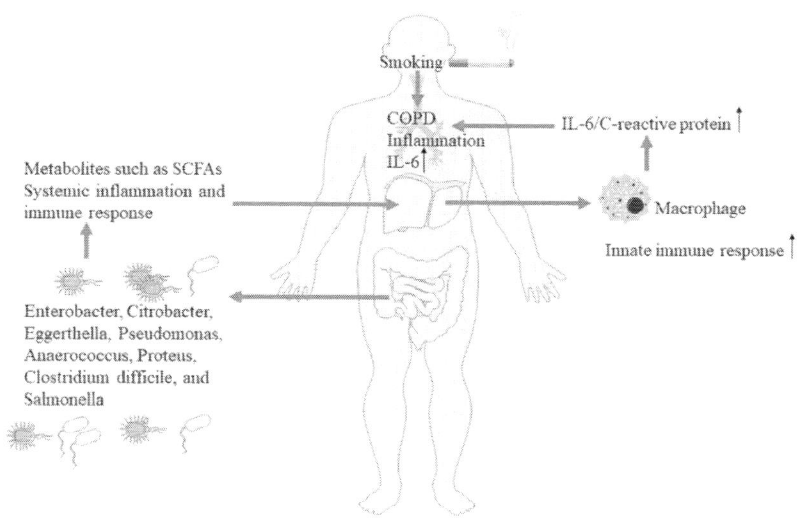

Figure 13: Above displays the involvement of the gut bacteria to stall COPD (2020 Zhang, Li, Wang, Tan, Zhang and Feng: *This is an open-access article distributed under the terms of the Creative Commons Attribution License (CC BY). The use, distribution or reproduction in other forums is permitted, provided the original author(s) and the copyright owner(s) are credited and that the original publication in this journal is cited, in accordance with accepted academic practice. No use, distribution or reproduction is permitted which does not comply with these terms*).

A 2020-published study explored the variations in the gut bacteria in COPD patients and healthy controls. The stool samples from 28 COPD patients and 29 healthy controls were evaluated by metagenomics and metabolomics. The COPD and healthy gut bacteria displayed a net difference in overall community composition. The former showed an increase in *Streptococcus* and *Rothia*, *Romboutsia*, *Intestinibacter*, and *Escherichia* while the levels of *Bacteroides*, *Roseburia* and *Lachnospira*, and *Ruminococcaceae* were lowered. The patients displayed an increase in the following families: *Bifidobacteriaceae*, *Eubacteriaceae*, *Lactobacillaceae*, *Micrococcaceae*, *Streptococcaceae* and *Veillonellaceae* while a dip was observed in *Desulfovibrionaceae*, *Gastranaerophilaceae* and *Selenomonadaceae*, *Bacilli* and *Clostridia*. The COPD and healthy controls showed a variation of 146 bacterial species with a predominance of *Rothia* and *Streptococcus* spp., *Romboutsia timonensis* and *Intestinibacter bartlettii* in the COPD patient guts while *Coprobacter fastidiosus* and *Coprobacter secundus*, *Rikenellaceae* genus *RC9* and *Christensenellales* family *CAG-74*) were seen more in the non-COPD individuals.

Further validation in 16 COPD patients and 22 controls further showed that COPD patients had an abundance of *Streptococcus* species: *S. parasanguinis_B* and *S. salivarius* further linked to higher glucosyltransferase activity. The transfer of these organisms, commonly the first colonizers of the mouth and gut from the oral microbiome is suggested to be the cause of the presence of these streptococci in COPD patient gut samples.

Interestingly, a review showed that the COPD patients consumed lower levels of fiber compared to the controls. Given that the importance of fiber in the microbes of the gut in serving as a substrate to influence the bacterial diversity and number to suppress immune inflammation, the intake of fiber for managing COPD is worth exploring.

Overall, the researchers outlined the distinctiveness of gut microbiota in COPD patients opening up the avenue for employing the bacteria of the gut as potential COPD markers.

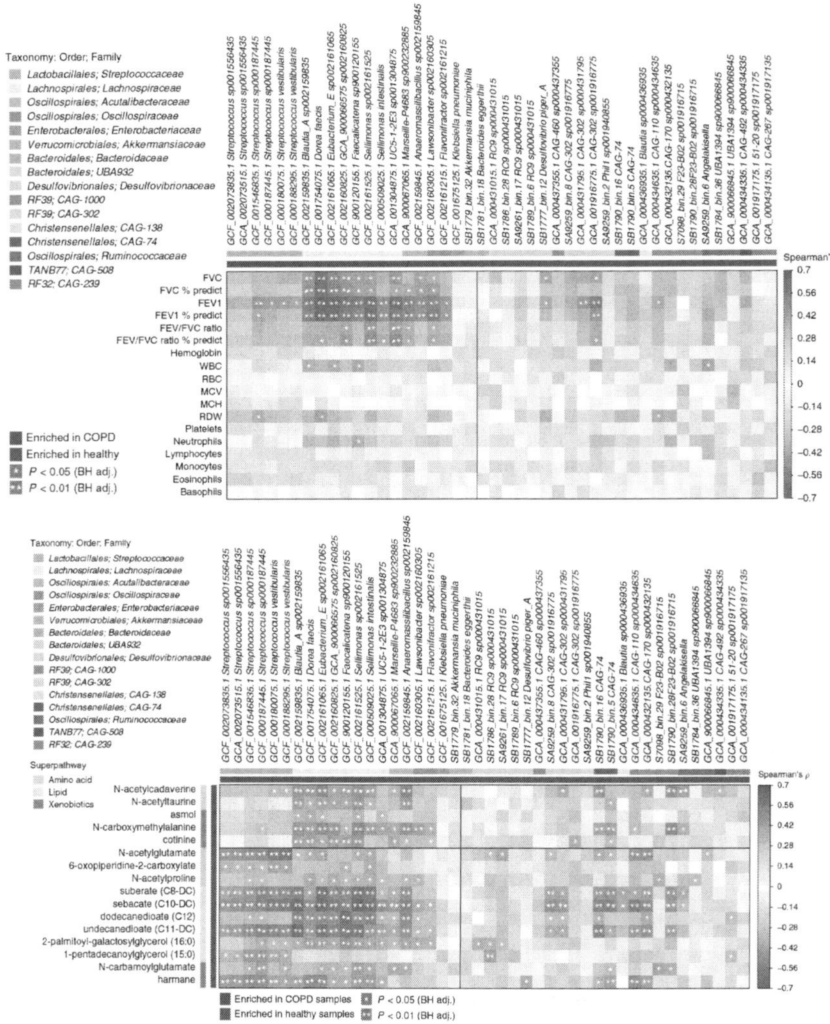

Figure 14: below shows ***TOP***: The connection of lung function and fecal microbiome; ***BOTTOM***: the differences in the species and metabolites with enrichment displayed at the left and top. White stars denote significant correlations. Colored bars represent the species order with family taxonomy and metabolite pathway (Bowerman et al, 2020: *Open Access: This article is licensed under a Creative Commons Attribution 4.0 International License, which permits use, sharing, adaptation, distribution and reproduction in any medium or format, as long as you give appropriate credit to the original author(s) and the source, provide a link to the Creative Commons license, and indicate if changes were made. The images or other third party material in this article are included in the article's Creative Commons license, unless indicated otherwise in a credit*

Key for the above: *FVC forced vital capacity, FEV forced expiratory volume, WBC white blood cell, RBC red blood cell, MCV mean corpuscular volume, MCH mean corpuscular haemoglobin, RDW red cell distribution width.*

2.6 OTHER VIRUSES AND FINAL WORDS

Our system is attacked by a slew of viruses; let us see how the gut microbiota is altered by few viral infections.

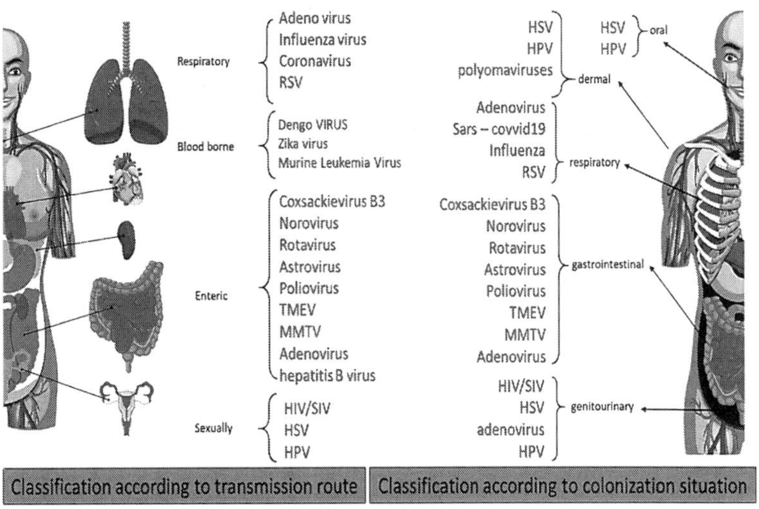

Figure 15: shows the common viruses targeting us (inclusive of the COVID-19 virus) (Baghbani et al, 2020: ***Open Access:*** *This article is licensed under a Creative Commons Attribution 4.0 International License, which permits use, sharing, adaptation, distribution and reproduction in any medium or format, as long as you give appropriate credit to the original author(s) and the source, provide a link to the Creative Commons licence, and indicate if changes were made. The images or other third party material in this article are included in the article's Creative Commons licence, unless indicated otherwise in a credit line to the material. If material is not included in the article's Creative Commons licence and your intended use is not permitted by statutory regulation or*

Acute gastroenteritis is caused by the Human Norovirus (HNoV) that has been associated with a lowering of Bacteroidetes in the gut. Another pathogen causing gastroenteritis is Rotavirus (RV). In infants administered an oral vaccine against RV, the levels of *Fusobacteria* and *Enterobacteriaceae were higher in those who didn't respond to the vaccine while Proteobacteria* and *Eggerthella were increased in the responders (*Baghbani et al, 2020).

Researchers Mutlu and team (2014) reported the next-generation sequencing-based analyses of the gut microbiome (terminal ileal and colonic mucosal areas) in 21 HIV positive and 22 control individuals. First, the HIV group displayed a lower operational taxonomic unit (OTU) count vs. the controls. Analyses revealed that the HIV patients had enrichment of "*Brachyspira, Campylobacter, Catenibacterium, Escherichia and other unclassified Enterobacteriaceae, unclassified Fusobacteriaceae, Mogibacterium, Prevotella, and Ralstonia*" while "*Akkermansia, Bacteroides, Blautia, Coprococcus, Dialister, Dorea, Faecalibacterium, Lachnospira, Roseburia, Ruminococcus, Odoribacter, Oscillospira genera, as well as unclassified bacteria from the following families, Barnesiellaceae, Lachnospiraceae, Peptostreptococcaceae, Rikenellaceae, and Ruminococcaceae*" were more in controls.

Looking at the bacteria, chronic diarrhea in HIV subjects is attributed to Brachyspira. Abscesses, infections, gastrointestinal infections, and UTIs are known to involve Enterobacteriaceae, Campylobacter, and Fusobacteriaceae. Uremia is linked to Catenibacterium while dental infections are associated with Mogibacterium. Prevotella despite being an inhabitant of the healthy gut is associated with lung and urinary tract infections. The plant pathogens Ralstonia are known to caused infections in immunocompromised individuals and cystic fibrosis patients.

The authors stated the lucid distinction of the gut microbiome of HIV patients and controls on the basis of the global microbiome composition suggestive of the crucial role of the GI tract microbiome as a target of HIV infection. This opens up the avenue of addressing HIV infections using pro/prebiotics to support the gut microbiome.

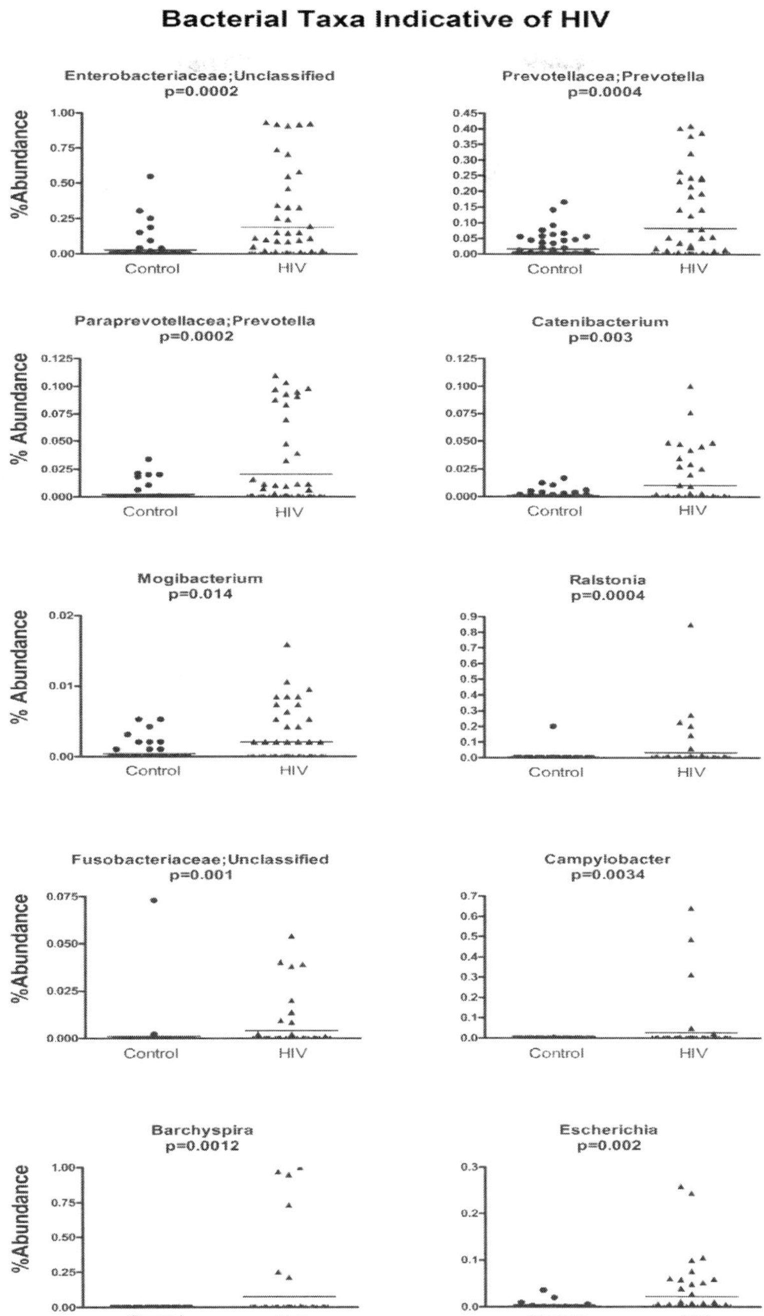
Bacterial Taxa Indicative of HIV
Enterobacteriaceae;Unclassified
p=0.0002
Prevotellacea;Prevotella
p=0.0004
Paraprevotellacea;Prevotella
p=0.0002
Catenibacterium
p=0.003
Mogibacterium
p=0.014
Ralstonia
p=0.0004
Fusobacteriaceae;Unclassified
p=0.001
Campylobacter
p=0.0034
Barchyspira
p=0.0012
Escherichia
p=0.002
% Abundance
Control
HIV

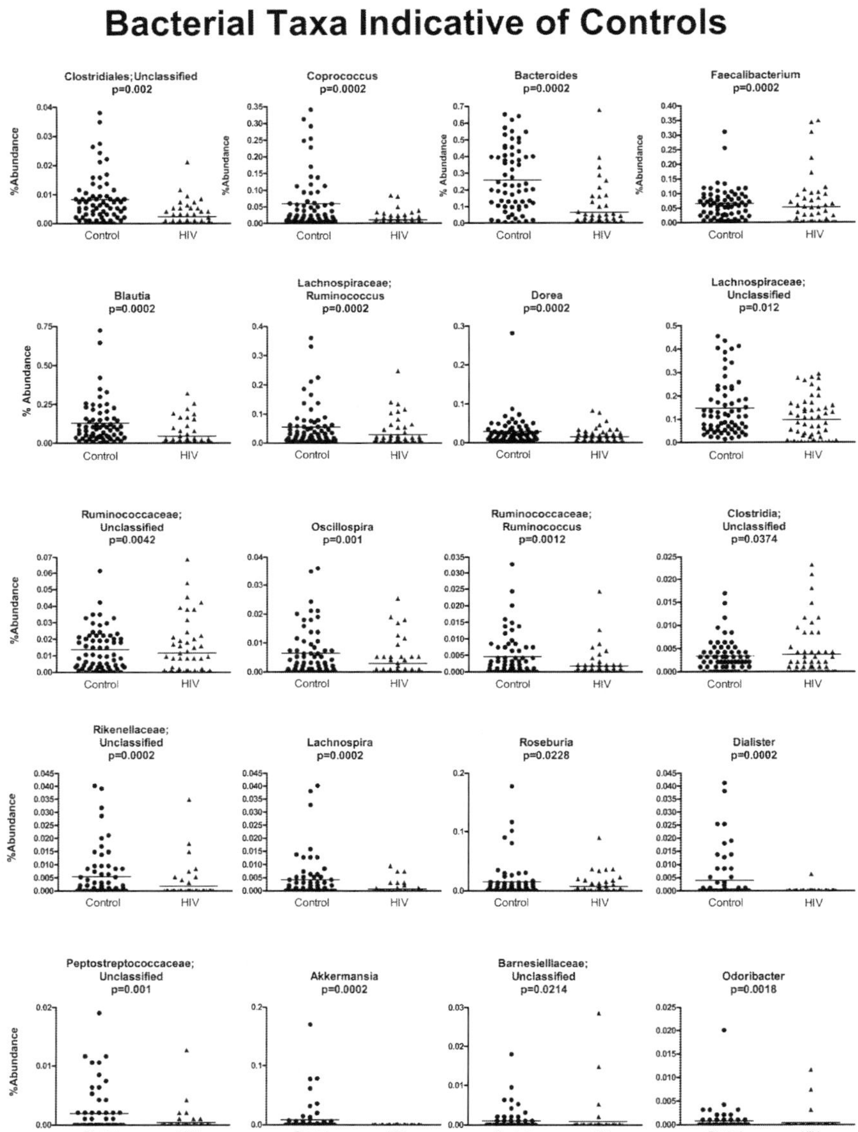

Figure 16: below shows ***TOP***: gut bacterial taxa that indicate HIV infection; ***BOTTOM***: as seen in controls; black dots represent controls and upward black triangles the HIV samples. The mean is shown as a horizontal line (Mutlu et al, 2014:*This is an open-access article distributed under the terms of the Creative Commons Attribution License, which permits unrestricted use, distribution, and reproduction in any medium, provided the original author and source are properly credited*):

In order to tackle the COVID-19 pandemic globally, the use of a face mask, regular hand washing, and social distancing, contact tracing, and quarantine have been proposed. Apart from the first line antivirals that are being administered and researched, many approaches look at addressing the inflammation induced by SARS-CoV employing immunomodulatory and anti-inflammatory therapies. Given that the diet and nutritional health can vitally impact overall health to prevent infections, can a healthy diet play a role in preventing the disease? Preliminary data look at the role of certain nutrients in the diet. These include zinc, vitamin C, and vitamin D that have benefic effects on the gut microbiota to overall impact immunity. A significant risk factor for COVID-19 is the inflammation seen in co-morbidities. This brings in the incorporation of food containing anti-inflammatory molecules such as flavonoids. While proof is yet to be established, not adopting unhealthy diets and following a healthy diet and lifestyle is also an important factor to prevent the disease (Zabetakis et al, 2020).

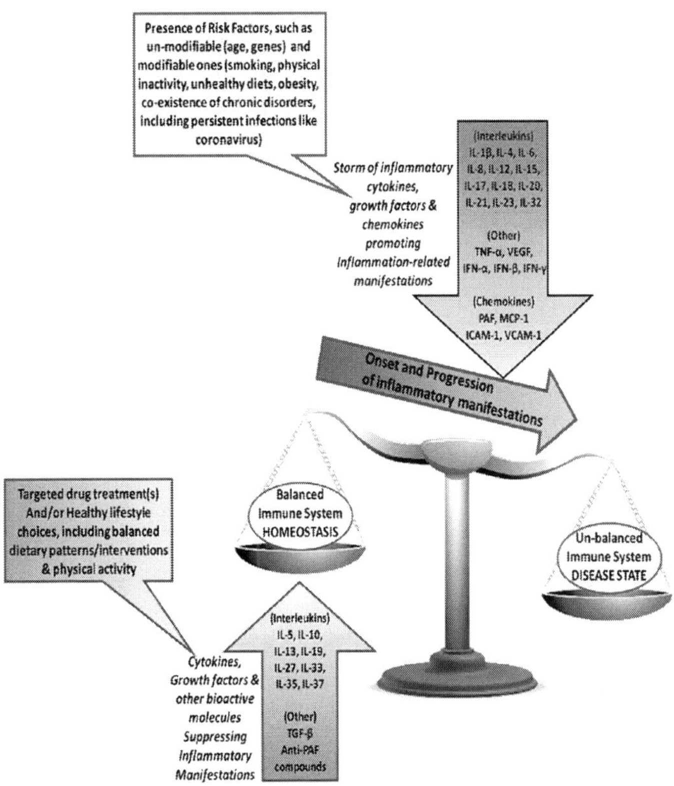

Figure 17: Above summarizes maintaining a healthy diet, gut, and lifestyle to quell the inflammatory molecules implicated in COVID-19 (Zabetakis et al,

2.7 REFERENCES

1. Zhang D, Li S, Wang N, Tan H-Y, Zhang Z and Feng Y (2020) The Cross-Talk Between Gut Microbiota and Lungs in Common Lung Diseases. *Front. Microbiol.* 11:301. doi: 10.3389/fmicb.2020.00301
2. Dharmage, S. C., Perret, J. L., & Custovic, A. (2019). Epidemiology of Asthma in Children and Adults. *Frontiers in pediatrics*, *7*, 246. https://doi.org/10.3389/fped.2019.00246
3. Network GA. The Global Asthma Report, Auckland, New Zealand. (2018).
4. Bisgaard H, Li N, Bonnelykke K, Chawes BL, Skov T, Paludan-Müller G, Stokholm J, Smith B, Krogfelt KA. Reduced diversity of the intestinal microbiota during infancy is associated with increased risk of allergic disease at school age. J Allergy Clin Immunol. 2011 Sep;128(3):646-52.e1-5. doi: 10.1016/j.jaci.2011.04.060. Epub 2011 Jul 22. PMID: 21782228.
5. Qian, LJ., Kang, SM., Xie, JL. *et al.* Early-life gut microbial colonization shapes Th1/Th2 balance in asthma model in BALB/c mice. *BMC Microbiol* **17,** 135 (2017). https://doi.org/10.1186/s12866-017-1044-0
6. Frati, F., Salvatori, C., Incorvaia, C., Bellucci, A., Di Cara, G., Marcucci, F., & Esposito, S. (2018). The Role of the Microbiome in Asthma: The Gut⁻Lung Axis. *International journal of molecular sciences*, *20*(1), 123. https://doi.org/10.3390/ijms20010123
7. Stokholm, J., Blaser, M.J., Thorsen, J. *et al.* Maturation of the gut microbiome and risk of asthma in childhood. *Nat Commun* **9,** 141 (2018). https://doi.org/10.1038/s41467-017-02573-2
8. Hufnagl, K., Pali-Schöll, I., Roth-Walter, F. et al. Dysbiosis of the gut and lung microbiome has a role in asthma. Semin Immunopathol 42, 75–93 (2020). https://doi.org/10.1007/s00281-019-00775-y
9. Chauhan S, Comprehensive review of coronavirus disease 2019 (COVID-19), Biomedical Journal, https:// doi.org/10.1016/j.bj.2020.05.023, ISSN 2319-4170, In press.]
10. Wiersinga WJ, Rhodes A, Cheng AC, Peacock SJ, Prescott HC. Pathophysiology, Transmission, Diagnosis, and Treatment of Coronavirus Disease 2019 (COVID-19): A Review [published online ahead of print, 2020 Jul 10]. JAMA. 2020;10.1001/jama.2020.12839. doi:10.1001/jama.2020.12839.

11. Singhal, T. A Review of Coronavirus Disease-2019 (COVID-19). Indian J Pediatr 87, 281–286 (2020). https://doi.org/10.1007/s12098-020-03263-6.

12. Shiv Sanjeevi Sripathi and Prerna Pandey (2021) An Insight Into The "Killers"-Viruses That Cause Epidemics. Delve Publishing; ISBN: 9781774078051.

13. Adhikari, S. P., Meng, S., Wu, Y. J., Mao, Y. P., Ye, R. X., Wang, Q. Z., Sun, C., Sylvia, S., Rozelle, S., Raat, H., & Zhou, H. (2020). Epidemiology, causes, clinical manifestation and diagnosis, prevention and control of coronavirus disease (COVID-19) during the early outbreak period: a scoping review. *Infectious diseases of poverty*, *9*(1), 29. https://doi.org/10.1186/s40249-020-00646-x

14. Donati Zeppa, S., Agostini, D., Piccoli, G., Stocchi, V., & Sestili, P. (2020). Gut Microbiota Status in COVID-19: An Unrecognized Player?. *Frontiers in cellular and infection microbiology*, *10*, 576551. https://doi.org/10.3389/fcimb.2020.576551

15. Segal JP, Mak JWY, Mullish BH, Alexander JL, Ng SC, Marchesi JR. The gut microbiome: an under-recognised contributor to the COVID-19 pandemic? *Therapeutic Advances in Gastroenterology*. January 2020. doi:10.1177/1756284820974914

16. Gu, S., Chen, Y., Wu, Z., Chen, Y., Gao, H., Lv, L., Guo, F., Zhang, X., Luo, R., Huang, C., Lu, H., Zheng, B., Zhang, J., Yan, R., Zhang, H., Jiang, H., Xu, Q., Guo, J., Gong, Y., Tang, L., … Li, L. (2020). Alterations of the Gut Microbiota in Patients with COVID-19 or H1N1 Influenza. *Clinical infectious diseases : an official publication of the Infectious Diseases Society of America*, ciaa709. Advance online publication. https://doi.org/10.1093/cid/ciaa709

17. Zuo, T, Zhang, F, Lui, GCY, et al. Alterations in gut microbiota of patients with COVID-19 during time of hospitalization. Gastroenterology. Epub ahead of print 20 May 2020. DOI: 10.1053/J.GASTRO.2020.05.048.

18. Gou W., Fu Y., Yue L., Chen G.-D., Cai X., Shuai M., et al. (2020). "Gut microbiota may underlie the predisposition of healthy individuals to COVID-19". *medRxiv.* doi: 10.1101/2020.04.22.20076091.

19. Trompette A, Gollwitzer ES, Yadava K, Sichelstiel AK, Sprenger N, Ngom-Bru C, Blanchard C, Junt T, Nicod LP, Harris NL, Marsland BJ. Gut microbiota metabolism of dietary fiber influences allergic airway disease and hematopoiesis. Nat Med. 2014 Feb;20(2):159-66. doi: 10.1038/nm.3444. Epub 2014 Jan 5. PMID: 24390308.

20. Ferreira C, Viana SD, Reis F. Gut Microbiota Dysbiosis–Immune Hyperresponse–Inflammation Triad in Coronavirus Disease 2019 (COVID-19): Impact of Pharmacological and Nutraceutical Approaches. *Microorganisms.* 2020; 8(10):1514. https://doi.org/10.3390/microorganisms8101514
21. McIlroy, J. R., Mullish, B. H., Goldenberg, S. D., Ianiro, G., & Marchesi, J. R. (2020). Intestinal microbiome transfer, a novel therapeutic strategy for COVID-19 induced hyperinflammation?: In reply to, 'COVID-19: Immunology and treatment options', Felsenstein, Herbert McNamara et al. 2020'. *Clinical immunology (Orlando, Fla.)*, *218*, 108542. https://doi.org/10.1016/j.clim.2020.108542
22. Wei, Y., Yang, J., Wang, J., Yang, Y., Huang, J., Gong, H., Cui, H., & Chen, D. (2016). Successful treatment with fecal microbiota transplantation in patients with multiple organ dysfunction syndrome and diarrhea following severe sepsis. *Critical care (London, England)*, *20*(1), 332. https://doi.org/10.1186/s13054-016-1491-2
23. Sanders, D. B., & Fink, A. K. (2016). Background and Epidemiology. *Pediatric clinics of North America*, *63*(4), 567–584. https://doi.org/10.1016/j.pcl.2016.04.001
24. Davis, P. B. Cystic fibrosis since 1938. *American journal of respiratory and critical care medicine* **173**, 475–482 (2006)
25. Schippa, S., Iebba, V., Santangelo, F., Gagliardi, A., De Biase, R. V., Stamato, A., Bertasi, S., Lucarelli, M., Conte, M. P., & Quattrucci, S. (2013). Cystic fibrosis transmembrane conductance regulator (CFTR) allelic variants relate to shifts in faecal microbiota of cystic fibrosis patients. *PloS one*, *8*(4), e61176. https://doi.org/10.1371/journal.pone.0061176
26. Fouhy, F., Ronan, N.J., O'Sullivan, O. *et al.* A pilot study demonstrating the altered gut microbiota functionality in stable adults with Cystic Fibrosis. *Sci Rep* **7,** 6685 (2017). https://doi.org/10.1038/s41598-017-06880-y
27. Koul P. A. (2013). Chronic obstructive pulmonary disease: Indian guidelines and the road ahead. *Lung India : official organ of Indian Chest Society*, *30*(3), 175–177. https://doi.org/10.4103/0970-2113.116233
28. Rabe, K. F. & Watz, H. Chronic obstructive pulmonary disease. *Lancet* **389**, 1931–1940 (2017).

29. Bowerman, K.L., Rehman, S.F., Vaughan, A. *et al.* Disease-associated gut microbiome and metabolome changes in patients with chronic obstructive pulmonary disease. *Nat Commun* **11,** 5886 (2020). https://doi.org/10.1038/s41467-020-19701-0
30. Baghbani, T., Nikzad, H., Azadbakht, J. *et al.* Dual and mutual interaction between microbiota and viral infections: a possible treat for COVID-19. *Microb Cell Fact* **19,** 217 (2020).https://doi.org/10.1186/s12934-020-01483-1
31. Mutlu, E. A., Keshavarzian, A., Losurdo, J., Swanson, G., Siewe, B., Forsyth, C., French, A., Demarais, P., Sun, Y., Koenig, L., Cox, S., Engen, P., Chakradeo, P., Abbasi, R., Gorenz, A., Burns, C., & Landay, A. (2014). A compositional look at the human gastrointestinal microbiome and immune activation parameters in HIV infected subjects. *PLoS pathogens*, *10*(2), e1003829. https://doi.org/10.1371/journal.ppat.1003829
32. Zabetakis I, Lordan R, Norton C, Tsoupras A. COVID-19: The Inflammation Link and the Role of Nutrition in Potential Mitigation. *Nutrients*. 2020;12(5):1466. Published 2020 May 19. doi:10.3390/nu12051466.

Chapter 3

THE GUT BACTERIA AND THE NERVOUS SYSTEM

"Your diet is a bank account. Good food choices are good investments"

-Bethenny Frankel

Shiv Sanjeevi Sripathi[1]

[1]Shiv Sanjeevi Sripathi secured second rank in Mumbai University and first rank in his college at his undergrad level and in his college during his Masters and has worked on a stem cell project at the Specialized Centre for Cell Based Therapy (SCCT), KEM Hospital at Mumbai, following which he was a Junior Research Fellow at Microbiology & Cell Biology Department, Indian Institute of Science, Bangalore from August 2009 to November 2011. The field of work involved cloning of cell wall genes and transcription factors in E. coli & M. smegmatis. This was followed by a stint as a Research and Development Executive at Transasia Biomedicals, Mumbai where he worked on patent writing and Immunoassays from 2011 till 2013. In 2013, he entered the teaching field where he has been tutoring biology and chemistry to the students of 11th and 12th grades along with English grammar and communication skills to people of all age groups. Shiv also loves helping people write scientific manuscripts and has co-edited several books on life sciences and aspects of biotechnology with Apple Academic Press.

3.1 THE GUT–BRAIN AXIS

दीपो भक्षयते ध्वान्तं कज्जलं च प्रसूयते |
यदन्नं भक्षयेन्नित्यं जायते तादृशी प्रजा ||

This ancient quote from India describes how just as how soot is produced by a lamp that "eats" (dispels) darkness, on the same lines one will create or produce as per his/her diet.

The phrase, "in disease and health" not only applies to wedding vows but also to the bidirectional association between the central nervous system (CNS) and the gut. While the CNS is influenced by sensory visceral signals from the gut to affect the mood and reflex; the biology of the gut is also influenced by the brain. The changes in the CNS status in response to the gut microbiota constitute the gut-brain axis.

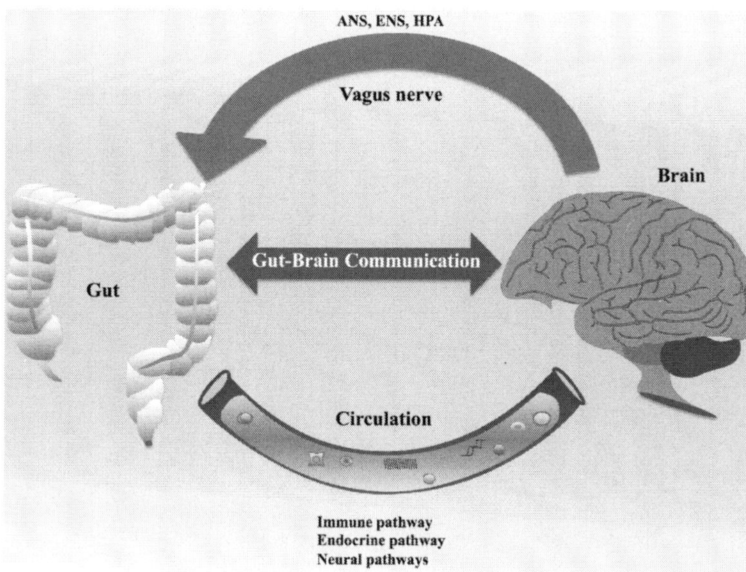

Figure 1: Above displays a scheme of the gut-brain axis (Suganya K and Koo, 2020: Licensee MDPI, Basel, Switzerland: *This article is an open access article distributed under the terms and conditions of the Creative Commons Attribution (CC BY) license (http://creativecommons.org/licenses/by/4.0/: You are free to: Share — copy and redistribute the material in any medium or format; Adapt — remix, transform, and build upon the material for any purpose, even commercially*).

The neuronal pathways are connected via afferent and efferent neurons: the former brings a signal in while the latter takes the signal out. The enteric nervous system (ENS) is composed of the myenteric and submucosal plexus that are involved in the regulation of secretion, peristalsis, and absorption in the gut. Other pathways involved are the autonomic nervous system (ANS), descending monoaminergic pathways, hypothalamic–pituitary–adrenal (HPA) axis, and the sympatho-adrenal axis.

The gut-brain signaling entails the role of intestinofugal neurons that facilitate the communication of the ENS with the CNS via the ganglia of the sympathetic nervous system (SNS). Among the multiple roles of the ANS such as controlling bile and carbohydrates, heartbeat, and breathing, the physiology of the gut can be impacted by the ANS directly via the CNS. The metabolites produced by the gut microbiota are sensed by the host cells to communicate with ANS synapses in the gut.

The direct and indirect associations of the gut microbiota along with their metabolites with CNS pathways entail the concept of "gut-brain axis". The manifestation of many CNS diseases involves gut microbiota disruption and dysbiosis (Ma et al, 2019; Suganya K and Koo, 2020).

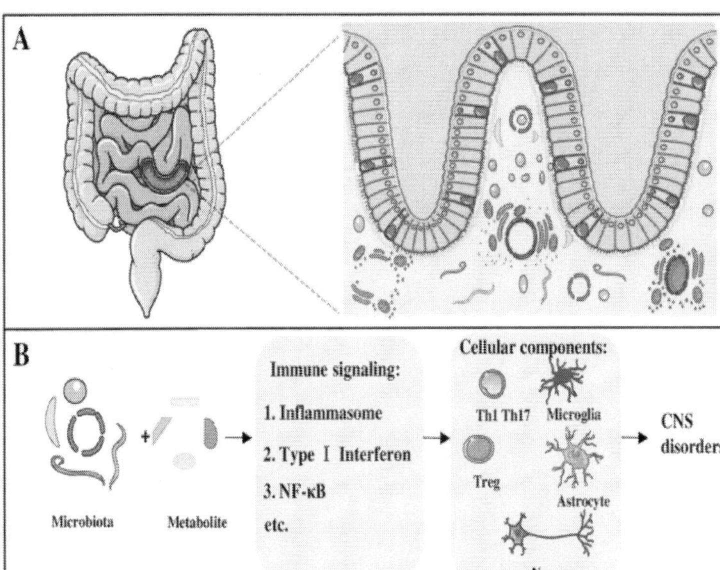

Figure 2: Above represents the interaction of the gut microbiota with the brain (Ma et al, 2019: *Open Access: This article is distributed under the terms of the Creative Commons Attribution 4.0 International License (http://creativecommons.org/licenses/by/4.0/), which permits unrestricted use, distribution, and re-*

production in any medium, provided you give appropriate credit to the original author(s) and the source, provide a link to the Creative Commons license, and indicate if changes were made. The Creative Commons Public Domain Dedication waiver (http://creativecommons.org/publicdomain/zero/1.0/) applies to the data made available in this article, unless otherwise stated).

3.2 THE GUT BACTERIA IN BRAIN DEVELOPMENT

The involvement of the microbiota in the developing gut in host functions was explored employing functional genomics and gnotobiotic mouse models. The approach involved the translocation of preterm human infant fecal homogenates in pregnant germ-free mice whose pups naturally acquired the microbiota via birth and nursing. The pups that received the sample from an infant showing low growth (defined as <10 gm/k/day weight gain: M_{PI}-L) and the infant with normal growth (>10 gm/k/day weight gain: M_{PI}-H) showed a difference with the latter showing a higher proportion of *Bacteroidetes* and *Actinobacteria*. It was further revealed that pathways associated with immune inflammation associated with NF-κB were activated in M_{PI}-L (from the host with poor growth). Interestingly, the inflammatory cytokine profile was seen at baseline in both the intestine and the serum showing that the **gut microbiota can influence the inflammatory pathways in the gut and in the system**. Given the occurrence of inflammatory conditions in preterm infants, such microbial communities that lower NF-κB activation can be vital to address such conditions (Lu et al, 2015).

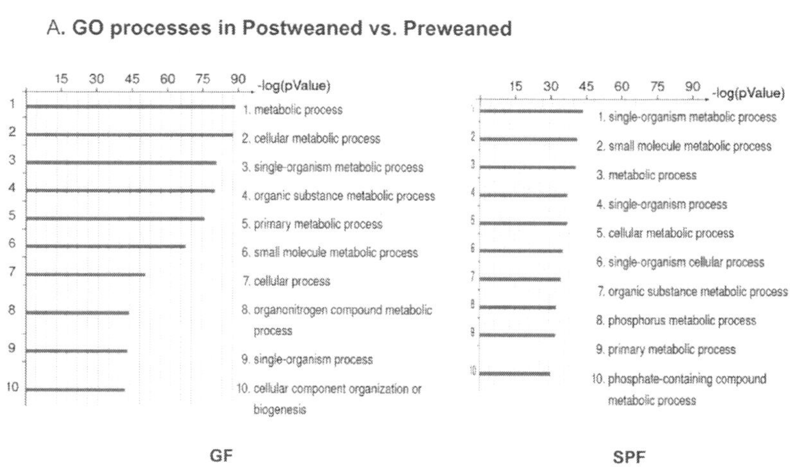

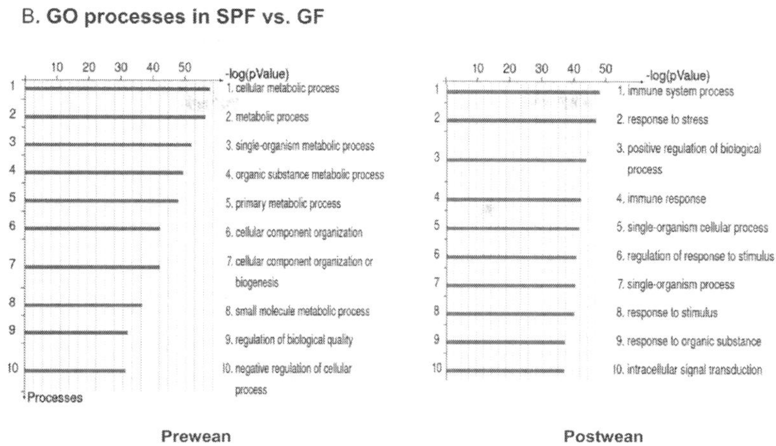

Figure 3: Above shows the microanalysis of A: the genes associated with development and B: genes influenced by the gut microbiota in the ileum of pre-weaned and post-weaned mice (2015 Lu et al: *This is an open-access article distributed under the terms of the Creative Commons Attribution License, which permits unrestricted use, distribution, and reproduction in any medium, provided the original author and source are credited*).

2018-published research by Lu and team assessed the role of the gut microbiota on brain development employing a germ-free mouse transfaunation model. Microbiota is known to cause high or low growth in infants was assessed for their effect on brain development. The colonization of newborns by microbes at and following birth that continues for up to 3 years influences the development of several organ systems inclusive of the brain. The risk of neurological conditions in preterm infants is known. Pregnant mice exposed to the human microbiome from normal growing infants and slow-growing infants were allowed to nurse their pups to allow the animals to develop the human microbiome.

The first markers assessed were NeuN (a marker of postmitotic mature neurons) and neurofilament proteins (NFs: that reflect axonal growth) to observe the effect of gut microbiota on neuronal development. The mice that received the M_{PI}-H sample (high-growth) displayed higher levels of NeuN and NF than the pups that received the M_{PI}-L (low-growth) sample. Further, delayed oligodendrocyte development and myelination were also observed in the M_{PI}-L mice as evidenced by higher levels of the markers NG2 and Olig2 in these animals vs. the M_{PI}-H group in the early development stage. Further, the former group also displayed lower levels of myelin basic protein

(MBP): a marker of myelination. This exciting observation is indicative of the **involvement of the gut microbiota on brain neuronal development**.

Given the vital involvement of the neurotransmitters glutamate (excitatory) and γ-aminobutyric acid (GABA: inhibitory) in the synaptic pathways of the brain, a quantitative RT-PCR-based assay of these molecules was done. The serotonin and dopamine pathways were lowered while that of the glutamatergic pathway was elevated in the microbiota from the low-growth infant. Further, animals with the microbiota from the preterm infant showing low growth displayed higher levels of IL-1β, TNF, and nitric oxide synthase (NOS1): all associated with inflammation and toxicity of neurons in the brain.

Additionally, the levels of IGF-1 were lowered in the liver and serum in the mice with microbes from the low-growth infant. This becomes interesting as neurogenesis (prenatal and early postnatal) involves the role of IGF-1. Overall, the researchers concluded that "*that colonization of different microbial communities influences brain development*" (Lu et al, 2018).

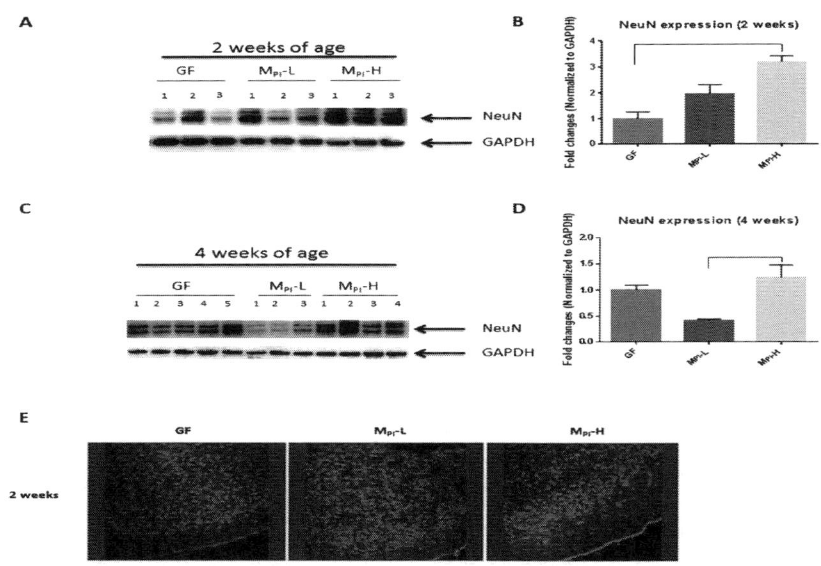

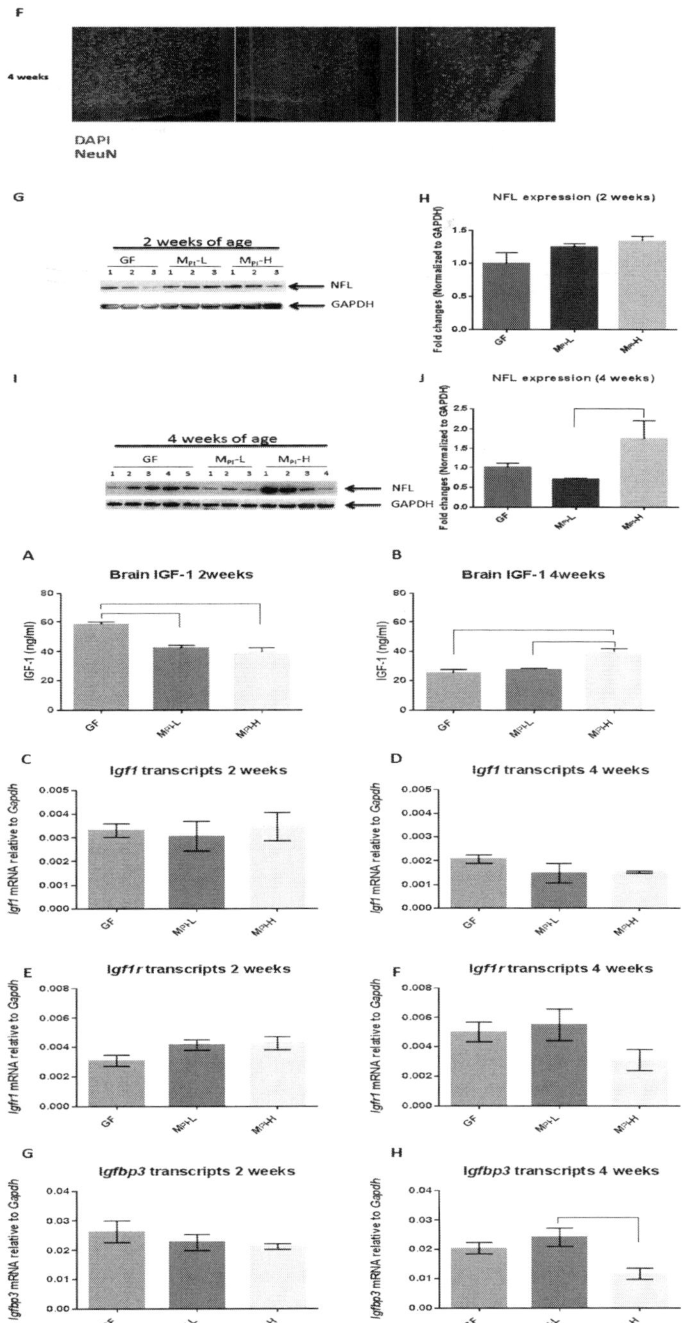

Figure 4: below displays the role of the gut microbiota in neuronal /brain development: ***TOP***: Gut microbiota influencing the level of the neuronal marker NeuN and neuronal development-western blot, immunohistochemistry and

quantitation of NeuN (A, B, C, and D); NeuN (Red) in DAPI (blue) immunofluorescence (E,F); NFL quantification (G, H, I, J); ***BOTTOM***: Lower levels of IGF-1 (a vital for neuron development), IGF1R (type 1 IGF receptor) and IGFBP3 (part of the IGF complex in serum) in the brain of the mice with poor growth microbiota (Lu et al, 2018, *Open-Access: This article is licensed under a Creative Commons Attribution 4.0 International License, which permits use, sharing, adaptation, distribution and reproduction in any medium or format, as long as you give appropriate credit to the original author(s) and the source, provide a link to the Creative Commons license, and indicate if changes were made. The images or other third party material in this article are included in the article's Creative Commons license, unless indicated otherwise in a credit line to the material. If material is not included in the article's Creative Commons license and your intended use is not permitted by statutory regulation or exceeds the permitted use, you will need to obtain permission directly from the copyright holder. To view a copy of this license, visit http://creativecommons.org/licenses/by/4.0/*):

3.3 GUT MICROBIOTA AND DEVELOPMENT OF THE MICROGLIA

The microglia are key cells in the central nervous system (CNS) described as macrophages residing in our brain. These cells exert a protective effect to protect from infection and injury. Aberrant functioning of these cells has been implicated in a range of diseases such as depression, Parkinson's disease (PD), and Alzheimer's disease (AD). These cells are vitally involved in the maintenance of CNS health and controlling cell survival and apoptosis. Interestingly, the first glial cells to migrate to the CNS are microglia to function in synapsis, neurogenesis, and excitotoxicity prevention.

Table 1: below shows how CNS health is maintained by the microglia (2018 Wang, Wang, Wang, Li, Jia, Song, Qin, Wang, Jin, Kitazato and Wang. *This is an open-access article distributed under the terms of the* ***Creative Commons Attribution License (CC BY)****. The use, distribution or reproduction in other forums is permitted, provided the original author(s) and the copyright owner(s) are credited and that the original publication in this journal is cited, in accordance with accepted academic practice. No use, distribution or reproduction is permitted which does not comply with these terms*)

Function	Microglia related process	Mediators	References
Neurogenesis	Guiding neurons and axons during the formation of neural circuits in prenatal development	-	(34)
	Integral components in neurogenic niches	CXCL12; CXCR4; ATP	(35, 36)
	Elimination of apoptotic neural stem cells and excess newborn progenitor cells	TAM; Gas6; Protein S	(37, 38)
	Age-associated reduction in neural stem cell proliferation	Proinflammatory cytokines secreted by microglia	(39)
Shaping Synapses	Synaptic pruning or connectivity	C1q; C3; CR3 (Microglia); CX3CR1 (Microglia); CX3CL1	(40, 41) (42, 43)
	Synaptic plasticity	Proinflammatory cytokines; ROS; NO; Neurotrophic factors; BDNF; IL-1β	(44, 45)
	Synaptic transmission	Glutamate; NADPH oxidase; NADPH receptor; PP2A; AMPA receptor;	(46)
	Synapse activity during neuropathic pain transmission	ATP; BDNF; Trk; KCC2 (Cation-chloride cotransporter KCC2)	(47)
Excitotoxicity prevention	Protection against NMDA-induced toxicity	ATP; P2X7; TNF-α	(48)

CXCL12, C-X-C motif chemokine 12; CXCR4, C-X-C chemokine receptor type 4; TAM, Tyro3, Axl, and Mertk receptors; Gas6, Growth arrest–specific 6; C1q, Complement component 1q; C3, Complement component 3; CR3, Complement receptor 3; CX3CR1, CX3C chemokine receptor 1; CX3CL1, C-X3-C motif chemokine ligand 1; PP2A, Protein phosphatase 2A; AMPA,α-amino-3-hydroxy-5-methyl-4-isoxazolepropionic acid; BDNF, Brain-derived neurotrophic factor; Trk, Neurotrophic receptor tyrosine kinase; P2X7, P2X purinoceptor 7; TNF-α, Tumor necrosis factor alpha.

Research suggests that microglia are influenced by commensal microbes probably do not involve the role of microbial-associated molecular patterns (MAMPs). Further, microglial maturation involves the role of short-chain fatty acids (SCFAs) that can go across the blood–brain barrier (BBB). In germ-free conditions, the activity of microglia was established by supplying SFCAs. The activity of nuclear factor-κB (NF-κB) is crucially controlled by SFCAs that in turn controls the microglial cell activity. The activity of microglia can be controlled by MAMPs or metabolites of the gut bacteria. Interestingly, the role of the vagus nerve in connecting the CNS activity by the gut has also been observed in preliminary research (Wang et al, 2018).

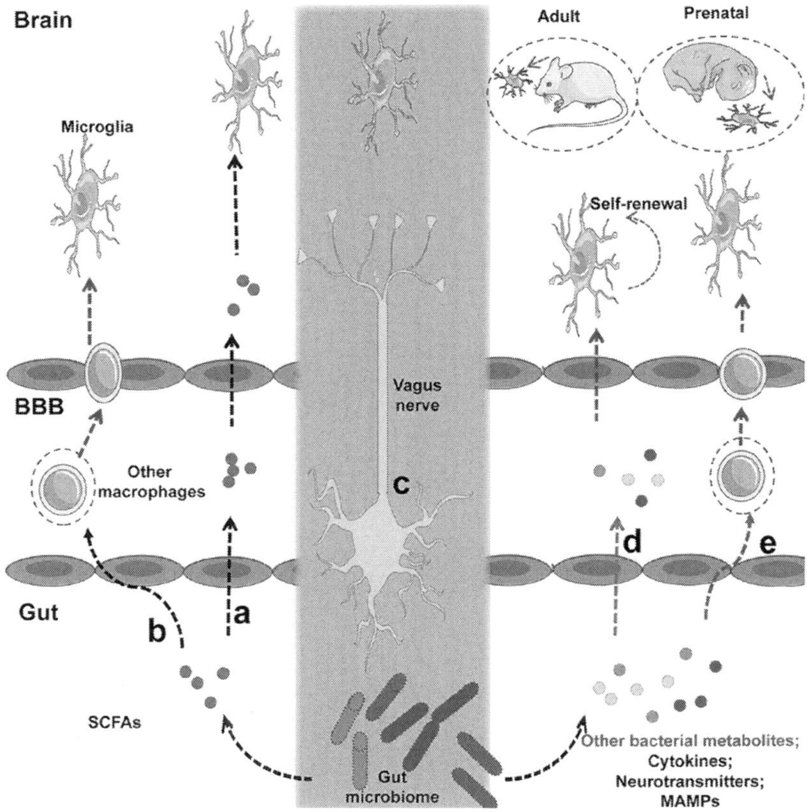

Figure 5: Above is illustrative of the gut microbes in influencing microglial functions : (a) Regulation of microglia by the SFCAs of gut bacteria (b) SFCA-recognizing receptor-bearing cells (c) The vagus nerve in possible direct control (d) MAMPs or metabolites from the gut (e) Gut flora influence peripheral macrophages to migrate across the BBB: What has been proven by research is shown in black lines while hypotheses are presented in red (2018 Wang, Wang, Wang, Li, Jia, Song, Qin, Wang, Jin, Kitazato and Wang. *This is an open-access article distributed under the terms of the* ***Creative Commons Attribution License (CC BY)****. The use, distribution or reproduction in other forums is permitted, provided the original author(s) and the copyright owner(s) are credited and that the original publication in this journal is cited, in accordance with accepted academic practice. No use, distribution or reproduction is permitted which does not comply with these terms*).

3.4 CHANGING THE BRAIN STRUCTURE?

"Tell me what you eat, and I will tell you who you are"

-Jean Anthelme Brillat-Savarin

Research has shown that the microbial community structure along with its gene expression was conspicuously altered with a short term diet alteration of five days (David et al, 2014). An interesting article by researchers Ong and team (2018) reported the changes in the structures of neural tissue in response to an alteration of the gut microbiome. This entailed the use of Sprague-Dawley rats exposed to a control diet with fats accounting for 17% of the energy; high-fat diet with fats accounting for 60% of the energy; a high-fiber-low energy diet with 40% fiber, and a high-protein-low carbohydrate diet with proteins accounting for 75% of the energy and carbohydrate for 6%. Along with diffusion tensor imaging (DTI) of the brain samples, 16S rRNA sequencing of the microbiome was done.

Analyses of the DTI revealed that the control and high-fat diet groups displayed evident alterations in RD, AD, and TR (radial diffusivity, axial diffusivity, and trace: mean diffusivity [MD] $\times 3$, respectively). This was observed in the forebrain, the corpus callosum, neocortex, and the right external and left internal capsules. The RD and TR were lowered conspicuously in the protein diet group in the left frontal neocortex, the left external capsule, and the forebrain. By the end of three weeks of the diet, the changes in gut microbiome communities were specific to the diets administered.

Further analyses revealed that the high protein-low carbohydrate diet could be predicted by alterations in *L. garvieae* while for the high-fiber diet, this was Roseburia, and Barnesiellaceae.

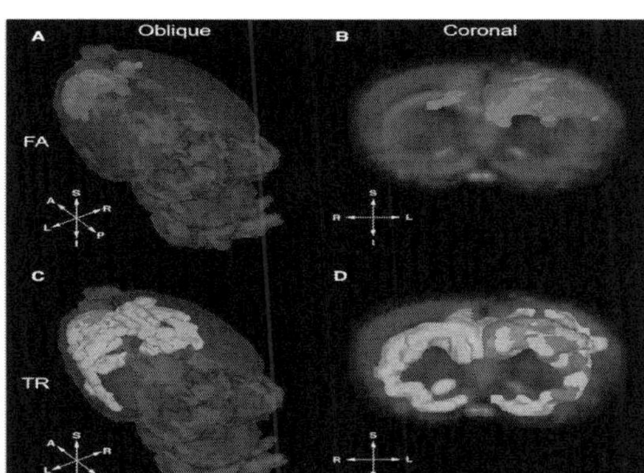

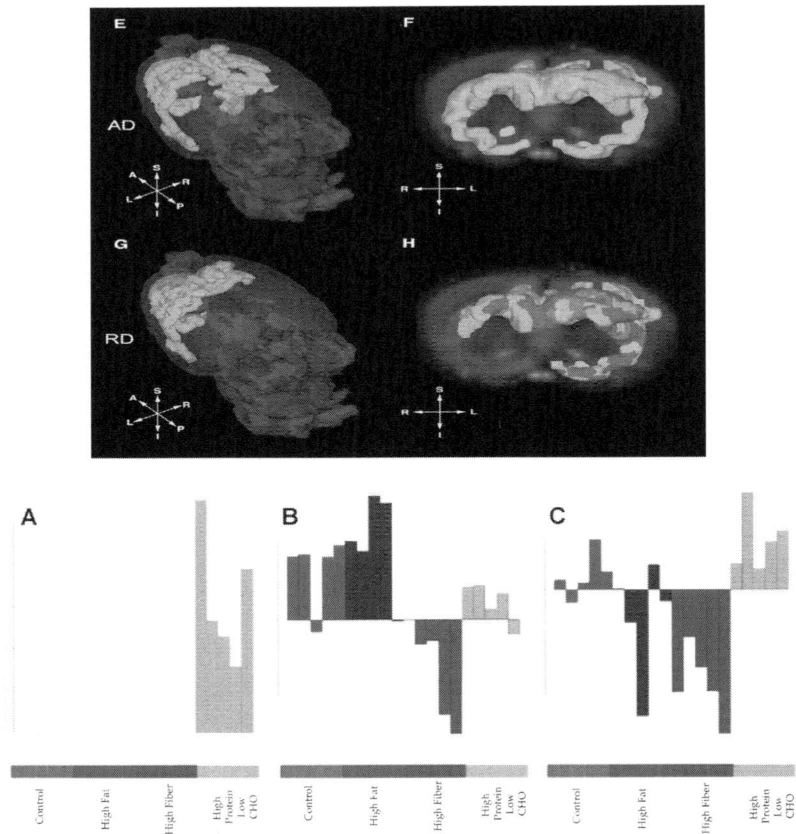

Figure 6: Above shows how the gut microbiota alterations can influence changes in specific areas of the brain; ***TOP***: Diffusion tensor imaging (DTI) analyses showing altered FA, TR, AD, or RD: the diet-specific changes are shown for high fat (yellow), high fiber (green), high protein (red) as per tract-based spatial statistics (TBSS) analysis that shows increased FA (a, b) and lowered trace (TR) (c, d) axial (AD) (e, f) and radial diffusivity (RD) (g, h); ***BOTTOM***: alterations in the abundance of (A) *L. garvieae* (B) Rosburia and (C) Barnesiellaceae in response to the diet (Ong et al, 2018: *Open-Access: This article is licensed under a Creative Commons Attribution 4.0 International License, which permits use, sharing, adaptation, distribution and reproduction in any medium or format, as long as you give appropriate credit to the original author(s) and the source, provide a link to the Creative Commons license, and indicate if changes were made. The images or other third party material in this article are included in the article's Creative Commons license, unless indicated otherwise in a credit line to the material. If material is not included in the article's Creative Commons license and your intended use is not permitted by statutory regulation or*

exceeds the permitted use, you will need to obtain permission directly from the copyright holder. To view a copy of this license, visit http://creativecommons.org/licenses/by/4.0/).

Overall, changes in the brain structure due to changes in the gut microbiome were revealed by machine learning methods. Further, the alterations in the FA (fractional anisotropy) observed i.e. higher clustering in the left frontal neocortex of the fiber diet group and in the protein diet groups can putatively involve aberrant myelination or alterations in the dimensions of the axons, according to the scientists.

Unique gut bacterial populations were identified that could be associated with these alterations in the brain. An example is Roseburia: this becomes interesting as there are reports of this genus to alter the mood. Li and team (2016) utilized high-throughput sequencing to look at the effect of gut microbiota on the mood swing of healthy adults. These approaches were performed in a closed human life support system for 105 days in healthy adults. A negative mood was associated with Faecalibacterium, Bifidobacterium, Bacteroides, Parabacteroides, and Anaerostipes while a positive mood was associated with Roseburia, Phascolarctobacterium, Lachnospira, and Prevotella.

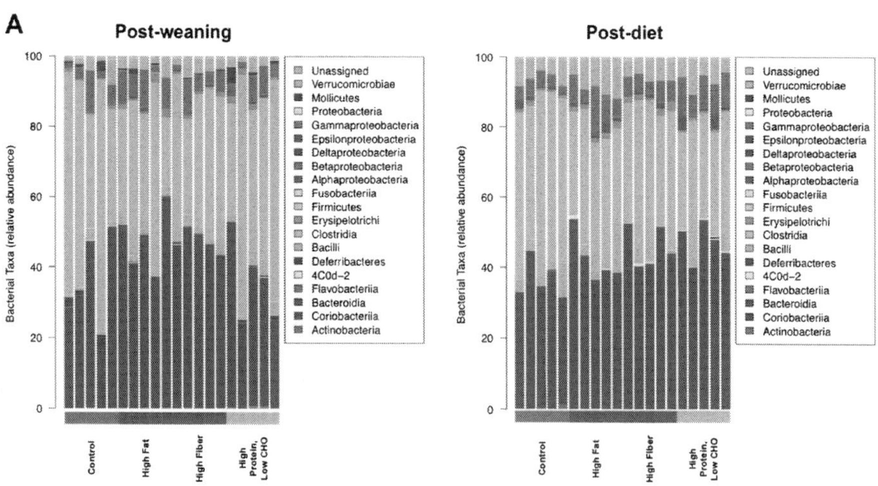

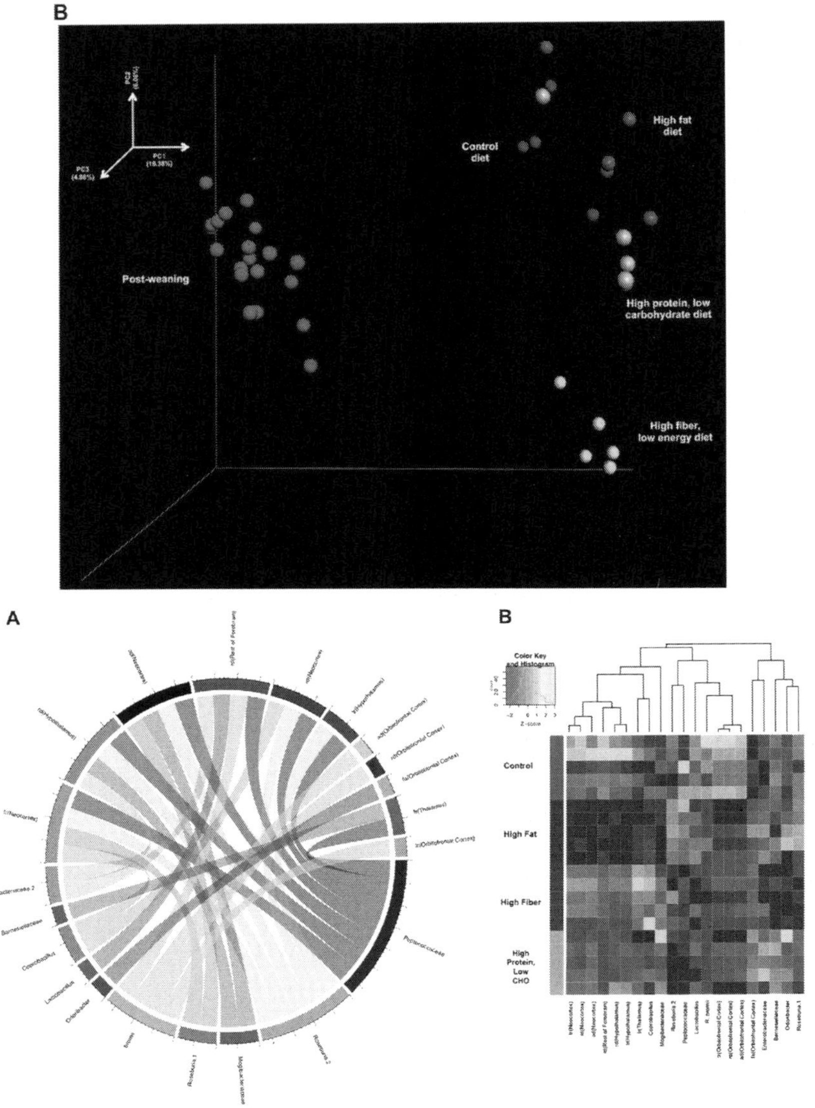

Figure 7: Above shows ***TOP***: abundance of the bacterial species (A), 16S rRNA analyses of the samples from the diets viz. blue for control, green for high-fat diet, pink for high protein- low carbohydrate, and yellow for the high-fiber diet (B); ***BOTTOM***: the gut microbiome bacteria that are associated with or predict the changes in the brain shown as (A) Circos plot (B) = hierarchically clustered heatmap*: fa fractional anisotropy, tr trace, rd radial diffusivity, ad axial diffusivity* (Ong et al, 2018: *Open-Access: This article is licensed under a Creative Commons Attribution 4.0 International License, which permits use, sharing, adaptation, distribution and reproduction in any medium or format, as long as*

3.5 A CHANGE IN GUT MICROBES CHANGE WITH ASD AND A SOLUTION:

"How we do food is how we do life. Every meal is a metaphor for how you show up in the world. Are you present? Are you complaining? Are you multitasking? Add love, celebration, time, communion, and gratitude to every meal and make every meal the best meal ever."

-Marc David & Emily Rosen

According to Park et al (2016), the Greek word αὐτός (autos) accounts for the origin of the word autism. A set of neurodevelopmental disorders is autism spectrum disorder (ASD) with compromised communication and social interactions. The prevalence is around 1-2% that is on the rise. The condition involves a complex interaction of genetics and the environment.

The gut microbiota was assessed in children with autism (AD), Pervasive Developmental Disorder Not Otherwise Specified (PDD-NOS) vs. that of healthy children by De Angelis and team (2013). The study was conducted in 10 children with PDD-NOS, AD, and with no known diseases (healthy children: HC), as per DSM-IV-TR criteria and analyses by Autism Diagnostic Interview-Revised (ADI-R), Autistic Diagnostic Observation Schedule (ADOS) and Childhood Autism Rating Scale (CARS). Following fecal sample extraction, tag-encoded FLX amplicon pyrosequencing (bTEFAP) was done and analyzed.

The majority sequences identified belonged to Firmicutes, Bacteroidetes, Proteobacteria, Actinobacteria, Fusobacteria and Verrucomicrobia. AD children showed lower levels of Firmicutes, Verrucomicrobia and Fusobacteria while that of Bacteroidetes were higher compared to the healthy ones.

Table 2: below illustrates the major gut bacteria that are metabolically active in the 3 groups of children studies: Pervasive Developmental Disorder Not Otherwise Specified (PDD-NOS), autistic (AD), and healthy (HC) (2013 De Angelis et al: *This is an open-access article distributed under the terms of the Creative Commons Attribution License, which permits unrestricted use, distribution, and reproduction in any medium, provided the original author and source are credited*)

Phylum/Family	Genus	Species	Avg (%) PDD-NOS	Avg (%) AD	Avg (%) HC	p-Val PDD-NOS vs HC	p-Val AD vs HC	p-Val PDD-NOS vs AD
Firmicutes/Ruminococcaceae	*Faecalibacterium*	*Faecalibacterium prausnitzii*	17.82	11.37	12.59	0.002	0.04	0.008
		Faecalibacterium sp.	21.94	15.19	17.38	0.06	0.03	0.02
	Ruminococcus	*Ruminococcus* sp.	15.59	8.47	8.79	0.007	0.07	0.006
		Ruminococcus obeum	0.18	0.12	0.19	0.08	0.04	0.02
		Ruminococcus gnavus	0.03	0.13	0.11	0.009	0.08	0.007
	Anaerofilum	*Anaerofilum* sp.	0.46	0.24	0.07	0.001	0.003	0.008
	Oscillospira	*Oscillospira* sp.	0.91	0.77	1.06	0.004	0.002	0.04
	Sporobacter	*Sporobacter* sp.	0.18	0.11	0.17	0.07	0.05	0.04
		Sporobacter termitidis	0.18	0.22	0.39	0.02	0.04	0.06
	Subdoligranulum	*Subdoligranulum* sp.	0.63	0.79	1.78	0.009	0.006	0.05
Firmicutes/Clostridiaceae	*Caloramator*	*Caloramator* sp.	0.10	0.39	0.09	0.09	0.005	0.006
	Sarcina	*Sarcina ventriculi*	0.00	0.16	0.00	0.09	0.01	0.09
		Sarcina sp.	0.00	0.28	0.00	0.10	0.02	0.09
	Clostridium	*Clostridium* sp.	5.83	7.97	7.38	0.02	0.04	0.03
		Clostridium orbiscindens	0.04	0.11	0.05	0.09	0.03	0.02
		Clostridium aminophilum	0.07	0.34	0.08	0.09	0.03	0.02
		Clostridium asparagiforme	0.07	0.31	0.08	0.08	0.009	0.009
		Clostridium bolteae	0.32	1.16	1.06	0.008	0.06	0.009
		Clostridium nexile	0.13	0.15	0.21	0.04	0.04	0.07
		Clostridium symbiosum	0.05	0.18	0.05	0.09	0.03	0.04
		Clostridium methylpentosum	0.02	0.01	0.15	0.03	0.02	0.09
		Clostridium bartlettii	1.64	1.15	2.30	0.02	0.04	0.03
Firmicutes/Lachnospiraceae	*Roseburia*	*Roseburia* sp.	1.77	6.09	1.33	0.04	0.003	0.004

		Roseburia faecis	0.27	0.45	0.94	0.009	0.02	0.04	
		Roseburia inulinivorans	0.39	2.66	0.14	0.04	0.008	0.009	
		Roseburia hominis	0.02	0.09	0.11	0.04	0.06	0.05	
		Roseburia intestinalis	0.18	0.31	0.50	0.03	0.04	0.04	
	Dorea	*Dorea* sp.	0.52	0.63	0.35	0.04	0.03	0.06	
	Coprococcus	*Coprococcus* sp.	0.04	0.12	0.17	0.04	0.05	0.04	
		Coprococcus eutactus	0.04	0.22	0.25	0.02	0.07	0.03	
	Lachnospira	*Lachnospira pectinoschiza*	0.09	0.14	0.20	0.02	0.05	0.05	
Firmicutes/Eubacteriaceae	*Eubacterium*	*Eubacterium coprostanoligenes*	0.80	0.51	0.84	0.08	0.04	0.04	
		Eubacterium ventriosum	0.20	0.08	0.25	0.07	0.01	0.02	
		Eubacterium sp.	4.89	2.19	6.18	0.007	0.001	0.008	
		Eubacterium siraeum	0.08	1.66	0.08	0.08	0.009	0.007	
		Eubacterium eligens	0.08	0.07	0.66	0.02	0.03	0.07	
	Enterococcus	*Enterococcus casseliflavus*	0.17	0.00	0.01	0.03	0.09	0.02	
Firmicutes/Streptococcaceae	*Streptococcus*	*Streptococcus salivarius*	0.11	0.17	0.35	0.02	0.03	0.06	
		Streptococcus thermophilus	0.24	0.38	0.12	0.02	0.009	0.04	
		Streptococcus sp.	0.16	0.16	0.20	0.04	0.04	0.09	
Firmicutes/Erysipelotrichaceae	*Turicibacter*	*Turicibacter* sp.	0.06	0.07	0.53	0.007	0.006	0.08	
		Turicibacter sanguinis	0.33	0.18	0.13	0.03	0.07	0.03	
Bacteroidetes/ Bacteroidaceae	*Bacteroides*	*Bacteroides* sp.	9.67	14.56	12.66	0.03	0.04	0.02	
		Bacteroides uniformis	0.70	1.35	3.12	0.01	0.02	0.03	
		Bacteroides fragilis	0.16	0.61	0.04	0.02	0.008	0.009	
		Bacteroides massiliensis	0.50	0.07	0.18	0.02	0.03	0.009	
		Bacteroides faecis	0.00	0.33	0.05	0.05	0.02	0.01	
		Bacteroides ovatus	0.07	0.12	0.16	0.02	0.06	0.04	
		Bacteroides vulgatus	0.53	1.17	0.80	0.03	0.007	0.005	
		Bacteroides intestinalis	0.00	0.79	0.04	0.06	0.03	0.02	
		Bacteroides coprocola	1.29	0.00	0.35	0.01	0.03	0.008	
Bacteroidetes/ Porphyromonadaceae	*Barnesiella*	*Barnesiella intestinihominis*	0.25	0.74	0.54	0.02	0.04	0.008	

	Odoribacter	*Odoribacter splanchnicus*	0.11	0.21	0.17	0.06	0.05	0.04	
	Parabacteroides	*Parabacteroides* sp.	0.03	0.12	0.11	0.03	0.08	0.04	
Bacteroidetes/ Prevotellaceae	*Prevotella*	*Prevotella* sp.	0.00	0.37	0.89	0.01	0.02	0.01	
		Prevotella copri	0.00	0.54	0.37	0.009	0.04	0.006	
Bacteroidetes/ Rikenellaceae	*Alistipes*	*Alistipes* sp.	0.47	0.29	0.27	0.02	0.07	0.03	
		Alistipes onderdonkii	0.48	0.01	0.19	0.03	0.04	0.01	
		Alistipes shahii	0.42	0.25	0.17	0.03	0.04	0.04	
		Alistipes indistinctus	0.02	0.14	0.02	0.09	0.03	0.02	
		Alistipes putredinis	1.80	2.03	1.57	0.04	0.01	0.03	
Proteobacteria/ Sutterellaceae	*Parasutterella*	*Parasutterella excrementihominis*	0.09	0.15	0.10	0.09	0.04	0.04	
Proteobacteria/ Enterobacteriaceae	*Enterobacter*	*Enterobacter hormaechei*	0.04	0.18	0.06	0.06	0.03	0.03	
	Escherichia	*Escherichia coli*	0.09	0.06	0.16	0.04	0.03	0.06	
	Shigella	*Shigella* sp.	0.08	0.18	0.09	0.08	0.03	0.02	
Actinobacteria/ Bifidobacteriaceae	*Bifidobacterium*	*Bifidobacterium* sp.	0.31	0.10	0.31	0.10	0.03	0.03	
		Bifidobacterium adolescentis	0.28	0.09	0.23	0.06	0.01	0.009	
		Bifidobacterium catenulatum	0.21	0.08	0.11	0.04	0.06	0.03	
Actinobacteria/ Coriobacteriaceae	*Collinsella*	*Collinsella* sp.	0.15	0.01	0.06	0.03	0.06	0.02	
Fusobacteria/Fusobacteriaceae	*Fusobacterium*	*Fusobacterium* sp.	0.45	0.29	0.79	0.03	0.009	0.03	
Verrucomicrobia/ Verrucomicrobiaceae	*Akkermansia*	*Akkermansia muciniphila*	0.08	0.27	0.10	0.06	0.03	0.02	

AD children exhibited the highest microbial diversity, as revealed by the Chao and Shannon diversity index. PDD-NOS and HC children showed the maximum levels of *Faecalibacterium* and *Ruminococcus* while for AD children, this was inclusive of *Caloramator*, *Sarcina* and *Clostridium*. PDD-NOS or AD children also showed high levels of Bacteroidetes genera with AD showing also elevated Sutterellaceae and Enterobacteriaceae levels along with few species of *Alistipes* and *Akkermansia*. Further, AD patients had a lower level of *Bifidobacterium* species.

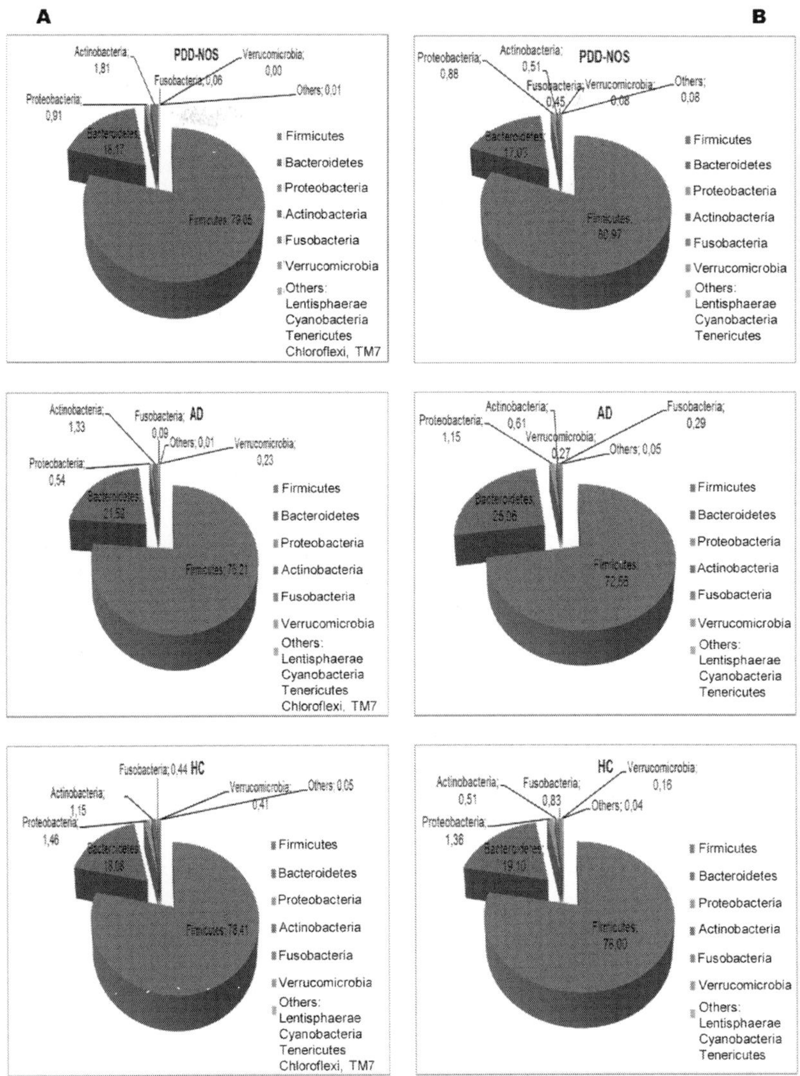

Figure 8: Above displays the bacteria in AD, PDD-NOS, and healthy children guts: (a) 16S-rDNA (B) 16S-rRNA: metabolic active bacteria (2013 De Angelis et al: *This is an open-access article distributed under the terms of the Creative Commons Attribution License, which permits unrestricted use, distribution, and reproduction in any medium, provided the original author and source are credited*).

AD children showed the highest abundance of Clostridiaceae: their ability to sporulate and synthesize toxic compounds like phenols have led to the hypothesis of the involvement of this group of bacteria in AD. Children with AD or GI inflammation often show the presence of high levels of Bacteroidetes, as seen in this study too. Further, the lower levels of *Bifidobacterium* become pertinent as these bacteria produce metabolites that are fermented upon by other intestinal microbes (De Angelis et al, 2013).

Though several approaches are utilized to address ASD ranging from educational and behavioral to pharmacological treatments, the modulation of gut bacteria too looks like a putative approach to help manage this condition. Given the gut microbiota dysbiosis observed in ASD, the utility of fecal microbiota transplant (FMT) in an open-label trial (ClinicalTrials.gov: NCT02504554) to address the challenges of ASD by researchers Kang and team (2017). As determined by the Autism Diagnostic Interview-Revised (ADI-R), 18 children with ASD were administered a 10-week Microbiota Transfer Therapy (MTT) treatment and a subsequent 8-week follow-up.

This system entailed an initial administration of vancomycin for 14 days to target pathogens in the gut followed by flushing of the bowels with MoviPrep to remove the antibiotic and the gut bacteria. A rectal or oral dose of Standardized Human Gut Microbiota (SHGM) was given: followed by maintenance doses. The controls were 20 children with no GI issues and were neurotypical).

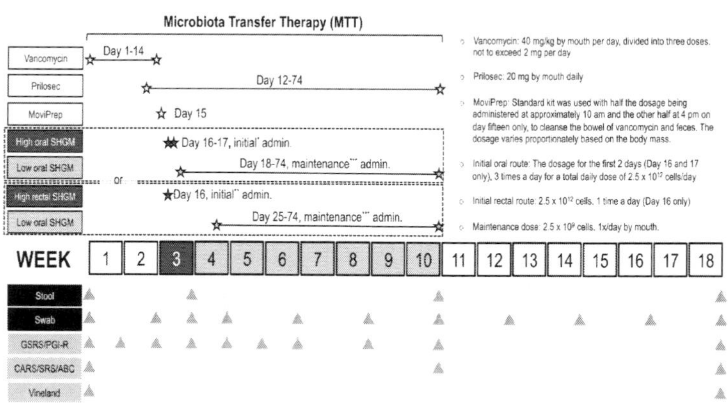

Figure 9: Above shows the approach utilized in the clinical trial using gut microbiota to address ASD symptoms: neurotypical in green and ASD group in purple (Kang et al, 2017: *Open-Access: This article is distributed under the terms of the Creative Commons Attribution 4.0 International License (http:// creativecommons.org/licenses/by/4.0/), which permits unrestricted use, distribution, and reproduction in any medium, provided you give appropriate credit to the original author(s) and the source, provide a link to the Creative Commons*

license, and indicate if changes were made. The Creative Commons Public Domain Dedication waiver (http://creativecommons.org/publicdomain/zero/1.0/) applies to the data made available in this article, unless otherwise stated).

The treatment resulted in a conspicuous lowering of GI symptoms associated with ASD as measured by a dip in the Gastrointestinal Symptom Rating Scale (GSRS) by 82% that was maintained after ceasing the treatment for 8 weeks. There were improvements in ASD symptoms as measured by the Parent Global Impressions-III (PGI-III), the Childhood Autism Rating Scale (CARS), the Social Responsiveness Scale (SRS), the Aberrant Behavior Checklist (ABC) and the Vineland Adaptive Behavior Scale-II (VABS-II). There was an increase in bacterial diversity in the children that reached a point where there was no difference in the diversity as seen in healthy controls from the initial scenario where the diversity was lesser in the ASD group. An extreme underrepresentation of *Bifidobacterium* initially in the ASD group increased by a factor of four and reached a level that was at par with that of the neurotypical children following MTT. The levels of the beneficial gut members *Desulfovibrio* and *Prevotella* also increased showing the shift in gut bacteria population in ASD patients to that of healthy children. In a nutshell, the safety of MTT along with its ability to shift the gut microbiota in a diseased condition to that seen in healthy children necessitates further research.

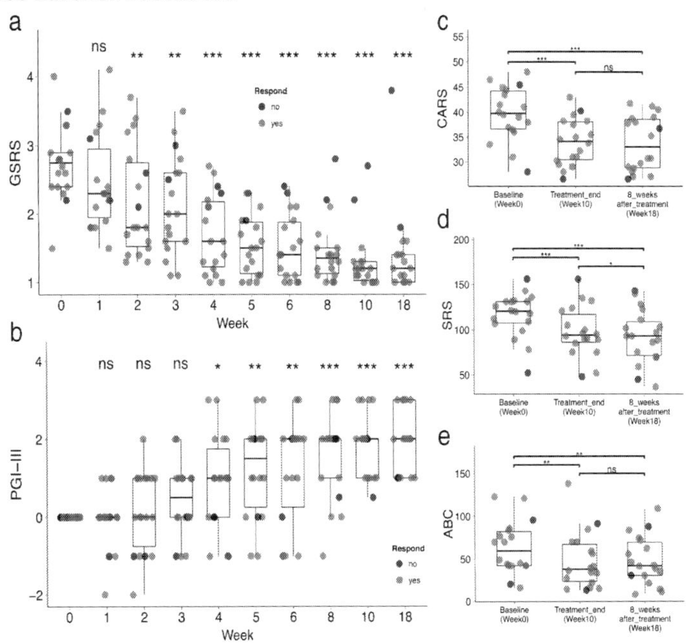

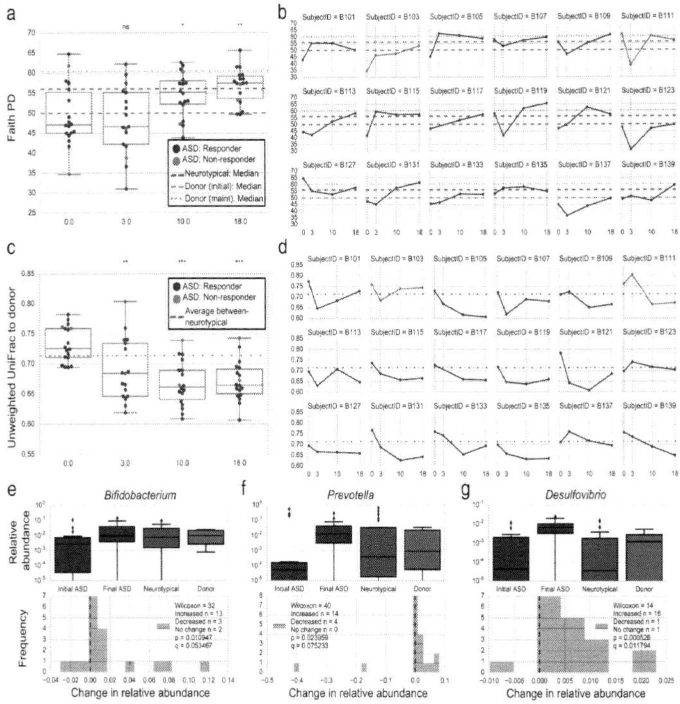

Figure 10: Above shows how the administration of a gut microbiota-based treatment approach resulted in ***TOP***: improved GI and ASD symptoms in ASD patients: a) GSRS scores b) PGI-III scores c) CARS d) Total SRS score e) ABC scores; Key: single asterisk indicates $p<0.05$, double asterisks indicate $p<0.01$, triple asterisks indicate $p<0.001$ and ns is not significant. The non-responders are in grey; ***BOTTOM***: a) Microbiota changes as per Faith's phylogenetic diversity (PD): donor (orange); maintenance (dotted); neurotypical (green) b) MTT recipient gut microbiota c) ASD gut microbiota UniFrac distances with the recipients and donors at par with each other d) Becoming similar to the donor microbiota e–g) Log-scale and abundances in Bifidobacterium, Prevotella, and Desulfovibrio (Kang et al, 2017: *Open-Access: This article is distributed under the terms of the Creative Commons Attribution 4.0 International License (http://creativecommons.org/licenses/by/4.0/), which permits unrestricted use, distribution, and reproduction in any medium, provided you give appropriate credit to the original author(s) and the source, provide a link to the Creative Commons license, and indicate if changes were made. The Creative Commons Public Domain Dedication waiver (http://creativecommons.org/publicdomain/zero/1.0/) applies to the data made available in this article, unless otherwise stated*).

3.6 GUT MICROBIOTA, THE BRAIN AND BEHAVIOR

The use of germ-free (GF) mice has allowed several insights into the roles of the gut microbiota. Given that mice are social like humans to explore social settings for pleasure, research by Desbonnet and team (2014) reported evidently impaired social habits in GF mice (particularly males) as observed in the three-chambered sociability test. This test (Crawley's sociability) looks at social affiliation and social memory in mice based on the time taken by a mouse in the compartments of three boxes with experimental indirect contact with a previously unfamiliar mouse. The time spent with another mouse rather than spending time alone is sociability while the time spent with an unknown mouse over a known one is the social novelty (Kaidanovich-Beilin et al, 2011).

Coming back to the experiment, GF mice did not display the tendency to spend time in a chamber with another mouse. Additionally, The GF mice did not spend the normally increased time to discover a new mouse. These observations are on the lines of what is reported for patients with neurodevelopmental disorders.

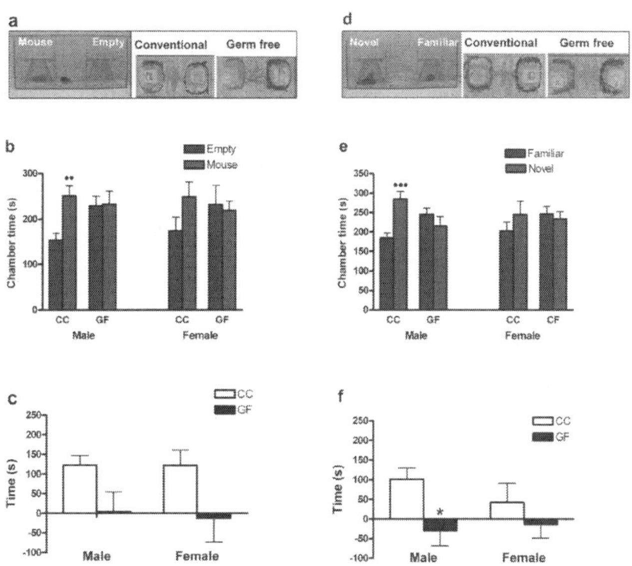

Figure 11: below shows the GF mice lacking gut microbiota show behavioral patterns similar to individuals with neurodevelopment issues; a to c represents lesser time spent with a mouse while d to f shows no increase in exploring new

social mates (Desbonnet et al, 2014: *This work is licensed under a Creative Commons Attribution 3.0 Unported License. To view a copy of this license, visit http://creativecommons.org/licenses/by/3.0/ : Free to Share — copy and redistribute the material in any medium or format; Adapt — remix, transform, and build upon the material for any purpose, even commercially*).

Following post-weaning bacterial colonization of the GF gut (GFC), social avoidance was reversed with no impact on the compromised social cognition reported above. This is indicative of the vital involvement of microbes in the social cognition aspects during pre-weaning. Further, GF mice indulged more in repeated self-grooming rather than social investigation that is similar to what is seen in autism. GF bacterial colonization reversed these effects showing for the first time **the role of the microbiota in behavior**. This becomes significant given that autism and schizophrenia are on the same lines as the observations of social avoidance and a higher occurrence in males (Desbonnet et al, 2014).

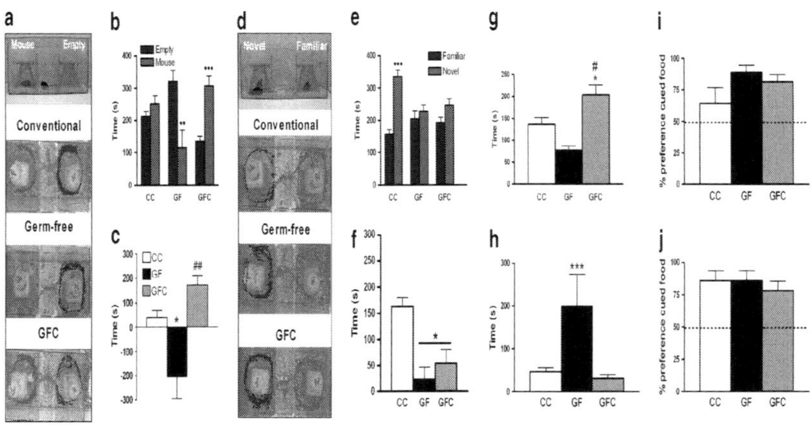

Figure 12: Above shows the impact of microbiota to reverse several aspects of behavior in GF mice; Tracking images of germ-free (GF) rearing and germ-free bacterial colonisation (GFC) animals in the three chambered test. (**a**), the time spent in each chamber (**b**) and the difference between time spent in mouse and empty chambers (**c**) reversing of social avoidance in GFC mice (**d**), the time spent in each chamber (**e**) and the difference between time spent in the chambers containing a novel and familiar mouse (**f**) novel food preference or the food to which a cage-mate was exposed before social interaction (**g**) and grooming time (**h**) during social interaction with demonstrator mice. There was no effect on the preference for cued food immediately after social interaction (**i**) and 24 h later (**j**) (Desbonnet et al, 2014: *This work is licensed under a Creative Commons Attribution 3.0 Unported License. To view a copy of this license, visit*

http://creativecommons.org/licenses/by/3.0/ : Free to Share — copy and redistribute the material in any medium or format; Adapt — remix, transform, and build upon the material for any purpose, even commercially).

Mar Gacias, Patrizia Casaccia and colleagues published an interesting report in eLife of how gut microbes impacted the brain and behavior of mice. Following the assignment of NOD (non-obese diabetic) mice into groups: one group received dH2O water, the second an antibiotic mix via a tube while the third and fourth groups received saline and antibiotic injections (Kuntz and Gilbert, 2016). The animals that were orally gavaged by dH2O for two weeks resulted in social avoidance with no effect on locomotion. This effect was not observed for parallel experiments with C57BL/6 mice. The antibiotic mix that was aimed to affect the gut microbiota (daily gastric gavage) did not cause this social avoidance in both NOD and C57BL/6 animals. The elevated plus maze (EPM) test revealed an anxiety-like behavior in NOD and C57BL/6 mice due to daily gavage. dH20 gavage resulted in despair-like behavior as seen in the forced swim test (FST) only in the NOD mice that was prevented by oral antibiotics. The researchers stated that oral antibiotic treatment did not affect the daily gavage induced anxiety suggestive of alterations in gut microbiota in mediating the depressive changes in the animals.

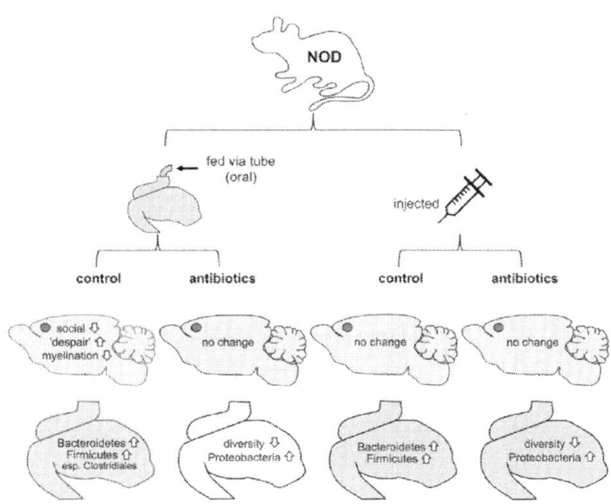

Figure 13: Above shows the influence of the gut microbes on the behavior of mice (Kuntz and Gilbert, 2016: *This article is distributed under the terms of the Creative Commons Attribution License, which permits unrestricted use and redistribution provided that the original author and source are credited*).

When the cecal and fecal samples were subjected to 16S rRNA sequencing analysis, antibiotic-treated animals displayed enrichment in Proteobacteria while Bacteroidetes and Firmicutes were enriched in animals treated with dH20 (termed as a vehicle). RNA sequencing of the medial prefrontal cortex (mPFC: linked to depression) of the dH20 gavaged mice that showed anxiety and antibiotic-treated mice revealed that the former group had lower expression of myelin as opposed to the latter group.

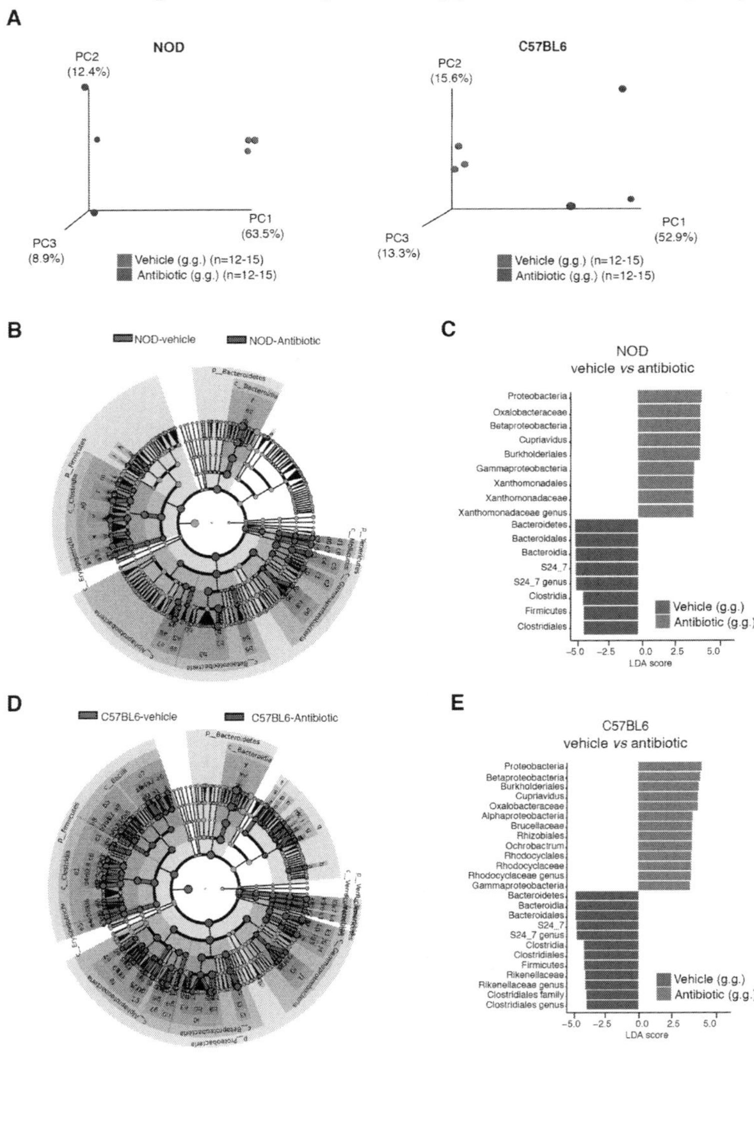

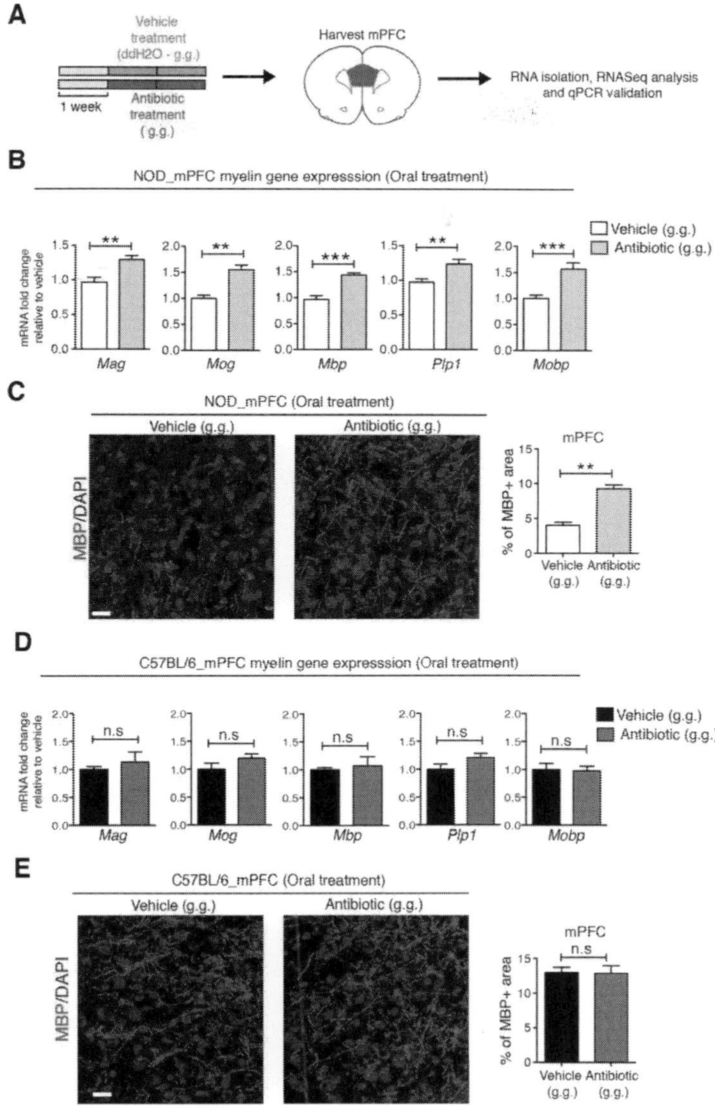

Figure 14: Above summarizes these findings. ***TOP***: changes in microbiota following treatment with dH20 or antibiotics; the microbiota are shown as dots in NOD (left) and C57BL/6 (right) mice (A); (B, D) Cladogram representing the most abundant taxa enriched in NOD treated with antibiotic (green) or vehicle (red) NOD (B) and C57BL/6 (D) mice. (C, E) Linear discriminant analysis (LDA) scores of microbiota abundance of NOD (C) and C57BL/6 (E) mice. ***BOTTOM***: lowered myelination-related transcripts in NOD mice subjected to 14 days of gavaging with dH20 (A) approach of 14 days of gastric gavage (g.g.) of vehicle or antibiotic. qPCR of myelin transcripts after vehicle or antibiotic

treatment of NOD (B) and C57BL/6 (D) mice. Confocal imaging of MBP+ (Myelin basic protein) fibers (red) against DAPI (blue) (C and E) (2016, Gacias et al; *This article is distributed under the terms of the Creative Commons Attribution License, which permits unrestricted use and redistribution provided that the original author and source are credited*).

Further, transplantation of the cecal content of vehicle-treated or antibiotic-treated NOD mice into C57BL/6 mice whose microbiota was eliminated via antibiotics revealed that the C57BL/6 recipients that received vehicle-gavaged (Group I) microbiota showed social avoidance. Analyses revealed that animals that showed depressive behaviors showed the presence of *Lachnospiraceae* and *Ruminococcaceae* of Clostridiales that were not detected in animals without the anxiety/depressive symptoms. These animals displayed lower levels of *Mag, Mog, Plp1, Mobp*genes associated with myelin genes while electron microscopy showed that these animals with social avoidance had lowered myelin thickness. Further, metabolomics revealed an increase in the level of cresol in these animals that inhibited myelin expression in *in vitro* oligodendrocyte cultures.

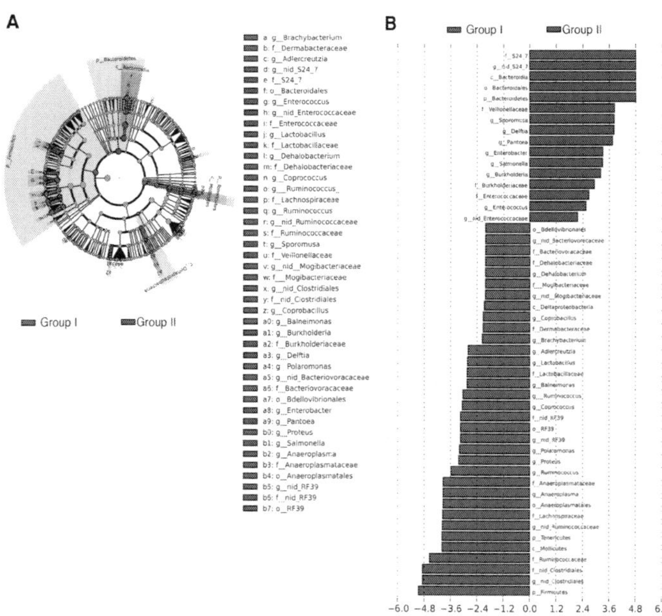

Figure 15: Above shows the bacterial taxa enriched in the recipients of cecal transplantation accounting for social avoidance shown as (A) Cladogram and (B) Linear discriminant analysis (LDA) scores of C57BL/6 recipients showing social avoidance (red) vs. those without social avoidance (green) (2016, Gacias

et al; *This article is distributed under the terms of the Creative Commons Attribution License, which permits unrestricted use and redistribution provided that the original author and source are credited*).

This work is indicative of the vital involvement of intestinal microbiota in the possible controlling of social behavior by altering the expression of metabolites.

3.7 THE BLOOD–BRAIN BARRIER AND THE GUT MICROBES

As we are aware of the importance of circulation and blood vessels, the vasculature changes with the tissue. The microvasculature of the CNS is described as the blood–brain barrier (BBB). This unique system is involved in the stringent control of molecules and cells between the brain and blood. This facilitates optimal functioning of the neurons while protecting them from toxins or damage (Daneman, and Prat, 2015).

To add to the growing findings of the gut–brain axis, researchers Hoyles and team (2018) reported an interesting role of propionate (a short-chain fatty acid [SCFA]) to communicate with the BBB. The treatment of hCMEC/D3 (human cerebromicrovascular endothelial cell line: a ubiquitous *in vitro* model of the BBB) and human brain tissue samples to propionate caused a change in the expression of 1136 genes with 203 genes associated with that of the BBB also detected. Of these genes, the most significant ones were eleven that were associated with according to the authors "*non-specific microbial infections and RNA- and endoplasmic reticulum-associated processes*". The exposure of the cell line to bacterial LPS was quelled by treating the cell line with propionate via suppression of the TLR-specific pathway. Further, another anti-oxidant associated pathway involving erythroid 2-like 2–NFE2L2 was also activated in these cells that lowered intracellular reactive oxygen species levels.

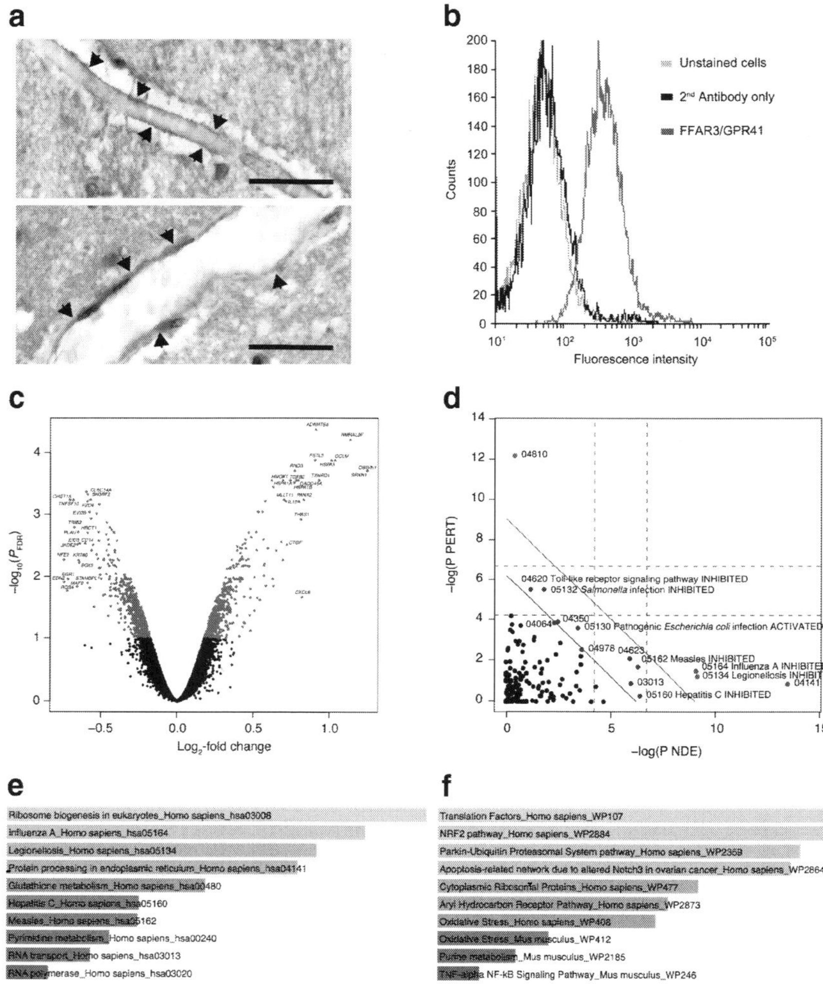

Figure 16: Above shows the effect of propionate (an SFCA) on the *in vitro* model of the BBB: Presence of FFAR3 (the propionate receptor) on a) human brain endothelium and b) hCMEC/D3 c) Differential gene expression d) Significant genes subjected to KEGG analyses Association with e) Enrichr analyses and f) Wikipathways: the result is more significant as the color gets lighter and the bars get longer; *pG. 04810. Regulation of actin cytoskeleton (inhibited); 04064, NF-kappa B signaling pathway (inhibited); 04978, mineral absorption (inhibited); 03013, RNA transport (activated); 04141, protein processing in the endoplasmic reticulum (activated); 04350, TGF-beta signaling pathway (activated); 04623, cytosolic DNA-sensing pathway (inhibited)* (Hoyles et al, 2018: Open-Access: *This article is distributed under the terms of the Creative Commons Attribution 4.0 International License (http://creativecommons.org/*

The protective effects of propionate: an SFCA produced by gut bacteria on the BBB shown in this nascent work becomes relevant as the BBB not only is the crucial link between the brain and blood but is also implicated in compromised cognition upon aberrant functioning. This opens up the avenue for future research to explore the roles of gut microbe-derived metabolites on the BBB to prevent inflammation or other pathologies.

3.8 DEPRESSION AND MICROBIOTA OF THE GUT: WHAT DOES RESEARCH SHOW?

The WHO estimates that 264 million people suffer from depression with the figures increasing (https://www.who.int/news-room/fact-sheets/detail/depression). Given this "depressing" trend, let us see select studies that have looked at the link between depression and the gut microbiota.

The gut microbiota of Flinders sensitive line (FSL) vs. that of Flinders resistant line (FRL) rats were assessed in a 2018-published article in the Psychopharmacology (Berl) journal. The former strain of animals displays altered behavior to serve as a model to explore depression. This was done given that the scientists observed a lowering of bacterial abundance and diversity in the guts of depressed patients in literature. The levels of *Firmicutes* and *Lachnospiraceae* were lowered while those of Proteobacteria and Actinobacteria were elevated. Further, the possibility of addressing depression by fecal microbiota transplantation (FMT) was also conducted.

Sequencing unveiled a lower number of OTUs or gut microbiota richness in FSL rats. Additionally, these models of depression showed decreased levels of *Elusimicrobia* and *Saccharibacteria*. The levels of *Rikenellaceae* , *Elusimicrobiaceae* , *Christensenellaceae* , *Peptostreptococcaceae*, *Bacteroidales S24-7 group*, and *Eubacteriaceae* were lower while those of *Bifidobacteriaceae*, *Lactobacillaceae* , *Lachnospiraceae* , *Veillonellaceae* , and *Helicobacteraceae* were more in the FSL rats. *Rikenellaceae* has been associated with fear and depression. Further, FSL rats also showed a lower profile of *Candidatus Saccharimonas*, *Alistipes*, and *Roseburia*: of which *Roseburia* is known to produce the health-promoting SFCA, butyrate.

Following FMT, the levels of *Rhodospirillaceae*, *Bifidobacteriaceae*, *Deferribacteraceae*, and *Peptostreptococcaceae* increased when the donor feces was from the FRL strain. The forced swim test revealed that the FSL rats were immobile and thus, "depressed" that was increased when the donor feces was from FSL while no effect was seen for FSR donor feces transplantation. Overall, the altered profile of gut microbiota in the depressed rats as opposed to the normal strain becomes significant to further delve into these aspects.

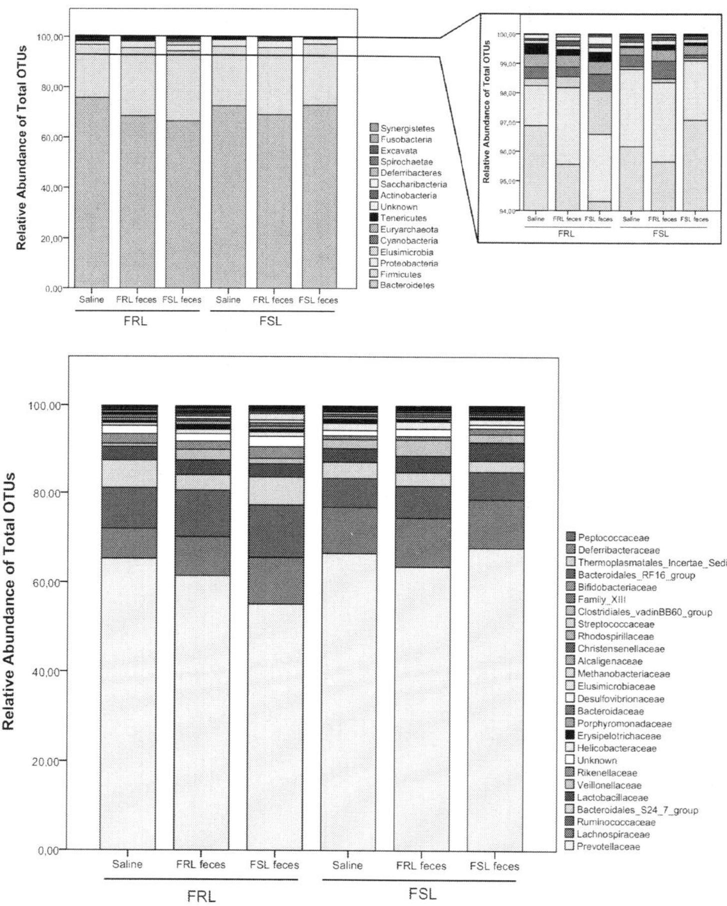

Figure 17: Above displays the abundance of the gut microbiota in FSL: Flinders sensitive line (depressed) and FRL: Flinders resistant line (normal) at the TOP: Phylum level; BOTTOM: family level (Tillmann et al, 2018, *Open-Access: This article is distributed under the terms of the Creative Commons Attribution 4.0 International License (http://creativecommons.org/licenses/by/4.0/), which permits unrestricted use, distribution, and reproduction in any medium, provid-*

ed you give appropriate credit to the original author(s) and the source, provide a link to the Creative Commons license, and indicate if changes were made).

2019-published research by Valles-Colomer and team evaluated the trend in depression and quality of life in a population of 1054 individuals of the Flemish Gut Flora Project. It was revealed that *Dialister* and *Coprococcus* spp were lowered in depression while a better quality of life was indicated by the levels of *Faecalibacterium* and *Coprococcus*. Interestingly, these latter two bacteria synthesize butyrate (an SFCA).

Analyses of a literature search by researchers Cheung, Goldenthal, Uhlemann, Mann, Miller and Sublettein the Frontiers in Psychiatry revealed that depressed patients and controls differed mainly in *Bacteroidetes, Firmicutes, Actinobacteria, Fusobacteria*, and *Protobacteria*. Overall, the levels of *Anaerostipes, Blautia, Clostridium, Klebsiella, Lachnospiraceae incertae sedis, Parabacteroides, Parasuterella, Phascolarctobacterium*, and *Streptococcus* were higher while that of *Bifidobacterium, Dialister, Escherichia/Shigella, Faecalibacterium*, and *Ruminococcus* were lower in depressed patients vs. controls assessed.

Table 3: below summarizes 6 studied analyzing the gut microbiome showing the taxa that were significant statistically in depressed vs. control individuals as presented: ↑, *increased in depression;* ↓, *decreased in depression;* ↔, *direction not deciphered; A-MDD, Active-MDD; R-MDD, recovering-MDD exhibited 50% reduction in Hamilton Depression Score after treatment. If "Jiang" is mentioned without identifying A-MDD or R-MDD, both exhibited the same directionality*: (© 2019 Cheung, Goldenthal, Uhlemann, Mann, Miller and Sublette: *This is an open-access article distributed under the terms of the Creative Commons Attribution License (CC BY). The use, distribution or reproduction in other forums is permitted, provided the original author(s) and the copyright owner(s) are credited and that the original publication in this journal is cited, in accordance with accepted academic practice. No use, distribution or reproduction is permitted which does not comply with these terms*)

Phylum	Class	Order	Family	Genus
Actinobacteria ↑Chen ↓Jiang	*Coriobacteridae*	*Coriobacteriales* ↑Zheng	*Coriobacteriaceae* ↑Zheng	
		Bifidonbacteriales	*Bifidobacteriaceae* ↑Chen	*Bifidobacterium* ↓Aizawa
		Actinomycetales	*Streptomycetaceae* ↑Chen	

			Actinomycetaceae ↑Chen	
			Nocardiaceae ↑Chen	
Bacteroidetes ↑Jiang ↑ Naseribafrouei ↓Chen ↓Lin	*Bacteroidia*	*Bacteroidales*	*Rikenellaceae* ↑Jiang ↔Chen	*Alistipes* ↑Jiang ↑Naseribafrouei ↓Zheng
			Prevotellaceae ↓Jiang (A-MDD) ↓Chen	*Prevotella* ↑Lin ↓Jiang
			Bacteroidaceae ↑Jiang (R-MDD) ↓Jiang (A-MDD)	*Bacteroides* ↑Jiang (R-MDD) ↓Jiang (A-MDD)
			Porphyromonadaceae ↑Jiang ↔Chen	*Parabacteroides* ↑Jiang
	Sphingobacteria	*Sphingobacteriales*	*Chitinophagaceae* ↓Chen	
			Marniabilaceae ↓Chen	
Firmicutes ↑Chen ↓Jiang ↑Lin	*Clostridia*	*Clostridiales*↑Zheng	*Lachnospiraceae* ↑Chen ↑Zheng ↓Jiang ↓Naseribafrouei	*Lachnospiraceae incertae sedis* ↑Jiang (A-MDD)
				Anaerostipes ↑Zheng
				Blautia ↑Jiang (A-MDD) ↑Zheng
				Roseburia ↑Jiang ↓Zheng
			Oscillospiraceae ↓Chen	*Oscillibacter* ↑Jiang (A-MDD) ↓Jiang (R-MDD) ↑Naseribafrouei
			Ruminococcaceae ↑Chen ↑Zheng ↓Jiang (R-MDD)	*Faecalibacterium* ↓Jiang ↓Zheng

				Ruminococcus ↓Jiang
			Clostridiaceae ↑Chen	*Clostridium* (XIX+IX) ↑Jiang (A-MDD)
				↑Lin
			Unclassified ↑Zheng	
	Negativicutes	*Selenomonadales*	*Veillonellaceae* ↓Jiang	*Megamonas* ↑Jiang (A-MDD) ↓Zheng
				Dialister ↓Jiang
			Acidaminococcaceae ↑Jiang ↓Zheng	*Phascolarctobac-terium* ↑Jiang
	Erysipelo-trichia	*Eerysipelotrichales*	*Erysipelotrichaceae* ↑Chen ↓Jiang (A-MDD) ↑Zheng	
	Bacilli	*Lactobacillus*	*Streptococcaceae*	*Streptococcus* ↑Lin
Fusobacteria ↑Jiang (A-MDD) ↓Jiang (R-MDD)	*Fusobacteriia*	*Fusobacteriales*	*Fusobacteriaceae* ↑Jiang (A-MDD)	
Proteobacteria ↑Jiang (A-MDD) ↓Chen ↓Jiang (R-MDD)	*Gammaproteo-bacteria*	*Enterobacteriales*	*Enterobacteriaceae* ↑Jiang ↓Chen	*Escherichia/Shi-gella* ↓Jiang (R-MDD)
				Klebsiella ↑Lin
	Betaproteobac-teria	*Burkholderiales*	*Sutterellaceae* ↓Chen	*Parasutterella* ↑Jiang (A-MDD)

Despite ongoing research highlighting novel aspects of depression, the condition is still a challenge. Given the involvement of altered gut microbiota suggested by studies, further research can help look at addressing this condition through the gut.

3.8 MICROBIAL METABOLITES AND THE NERVOUS SYSTEM

Several in vitro and in vivo assays revealed that the primary neuroinhibitor gamma-aminobutyric acid (GABA) is synthesized by *Escherichia* spp. and *Lactobacillus* spp, which are probiotic strains. The gut commensals produce short-chain fatty acids (SCFAs) that are associated with the G Protein-Coupled Receptor (GPR)- inflammasome to protect from colitis. SFCAs also modify histone by acetylating the promoter region to activate regulatory T-cells or Tregs, specifically Foxp3$^+$ Tregs. These Foxp3-expressing Tregs have been shown to secrete anti-inflammatory cytokines to quell the inflammation in experimental autoimmune encephalomyelitis (EAE) models. When SFCAs butyrate and propionic acid are produced in large amounts by the gut bacteria, the Foxp3$^+$ Treg levels are elevated to protect from inflammation. The possible involvement of the gut microbiota to control neurogenesis primarily in the hippocampus has been shown by research. GF mice display a lowered expression of occludin and claudin-5 that are vital in tight junctions making the BBB aberrant. This is attributed to the lack of gut microbes in them; when *Clostridium tyrobutyricum* (a butyrate producer) or microbial products were administered to these GF mice, the BBB structure was reestablished (Ma et al, 2019; Suganya K and Koo, 2020).

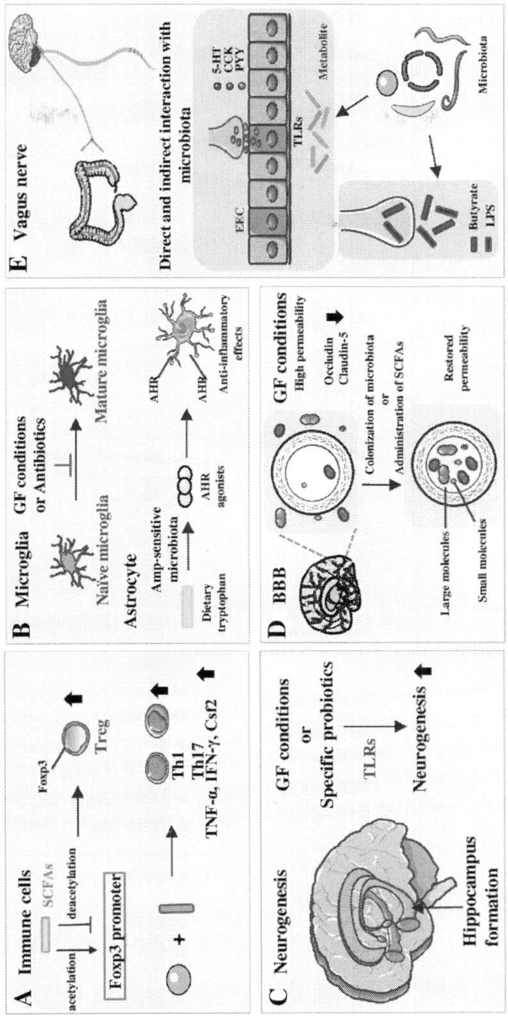

Figure 18: Above displays how various aspects associated with the CNS are linked to the gut microbiota. A) SFCA- Foxp3+ Tregs to target inflammation B) Changes in microglia by the gut microbiota C) GF/ antibiotic-treated mice show neurogenesis when the microbiota are removed D) SFCAs derived from gut microbiota could reverse the damaged orientation of the BBB (Ma et al, 2019: *Open Access: This article is distributed under the terms of the Creative Commons Attribution 4.0 International License (http://creativecommons.org/licenses/by/4.0/), which permits unrestricted use, distribution, and reproduction in any medium, provided you give appropriate credit to the original author(s) and the source, provide a link to the Creative Commons license, and indicate if changes were made. The Creative Commons Public Domain Dedication waiver*

(http://creativecommons.org/publicdomain/zero/1.0/) applies to the data made available in this article, unless otherwise stated).

The neurotransmitter glutamine is not only produced by the neurons but also by *Brevibacterium* spp, *L. lactis*, *L. plantarum*, and *Corynebacterium glutamicum* that enter the portal system via colonocytes. The gut microbes utilize tryptophan to produce Hydroxytryptamine (5-HT). GF mice with a mutated tryptophan hydroxylase 1 (TPH1: the enzyme that catalyzes this conversion) display a decrease in enteric neurons that was restored when wild-type mice microbes were administered. On the flip side, the activation of TLR in brain cells by LPS of the cell walls of few bacteria has shown to result in neuroinflammation (Ma et al, 2019; Suganya K and Koo, 2020).

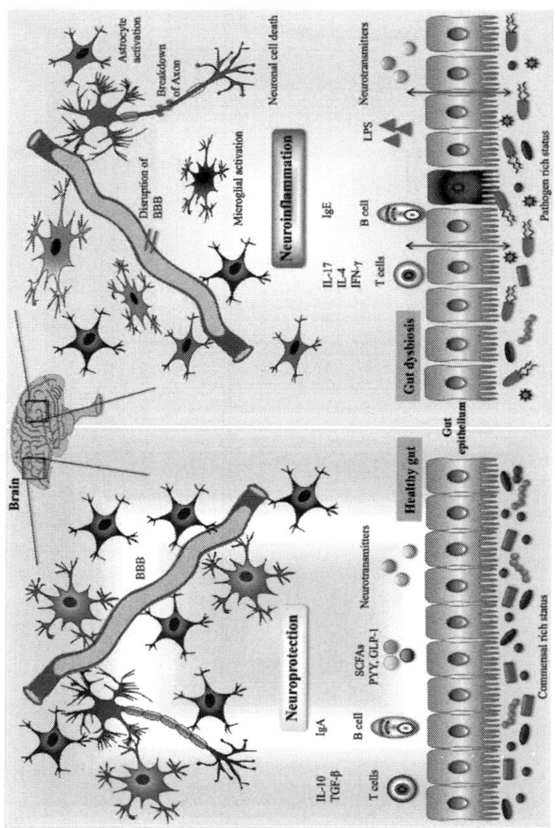

Figure 19: Above summarizes various molecules involved in the gut-brain axis in health (left) and disease (right). The various neurotransmitters and SFCAs produced by gut microbiota facilitate LPS movement to lower its level, protect the BBB while maintaining brain cell activity (left). However, dysbiosis results in increased lipopolysaccharides (LPS) or immune cell activation that affect the

nerve cells, the BBB to cause inflammation: Peptide YY (PYY), glucagon-like peptide-1 (GLP-1), and transforming growth factor-β (TGF-β) (Suganya K and Koo, 2020: Licensee MDPI, Basel, Switzerland: *This article is an open-access article distributed under the terms and conditions of the Creative Commons Attribution (CC BY) license (http://creativecommons.org/licenses/by/4.0/: You are free to Share — copy and redistribute the material in any medium or format; Adapt — remix, transform, and build upon the material for any purpose, even commercially*).

3.9 IN HEALTH AND IN DISEASE

The gut microflora of patients with multiple sclerosis (MS): a highly heterogeneous disease vs. that of controls was analyzed by Tremlett and Waubant (2018). The levels of Actinobacteria, associated with immune inflammation were up by a factor of 2.5 in MS patients. Further, an increased abundance of *Desulfovibrio* and *Bifidobacterium* associated with inflammation was also reported.

The microbiome of 108 elders was assessed: of these, 22.2% had Alzheimer's disease (AD), 30.6% had other causes of dementia and the rest showed no dementia. Shotgun metagenomics revealed that AD patients had a lowered abundance of *Lachnoclostridium* spp while the abundances of *Bacteroides* spp., *Alistipes* spp., *Odoribacter* spp., and *Barnesiella* spp. were more. The elders with other dementia showed a decrease in *Collinsella*, *Roseburia* , *Eubacterium* and *Lachnoclostridium* spp. it was further revealed that AD could be predicted by malnutrition, higher frailty and gut microbes that have been implicated in inflammation. Particularly, the levels of *Roseburia hominis*, *Eubacterium*, *Clostridium*, *Butyrivibrio* and *Faecalibacterium prausnitzii* were lower in these patients. Interestingly, these organisms are known producers of butyrate. The increased *Odoribacter splanchnicus* level in AD patients is not surprising given the involvement of this organism in mice models of AD. The other organism at higher abundance in AD seniors, *Bacteroides vulgatus* is known to cause neuroinflammation. Another organism, *Desulfovibrio fairfieldensis* was enriched in AD patients, an organism seen in several dysbiotic conditions.

Overall, it was shown that AD patients show enrichment of microbes involves in dysbiosis or inflammation along with a lowering of SFCA (butyrate)-producing organisms. This opens up an interesting avenue of research given the connection between the gut and AD (Haran et al, 2019).

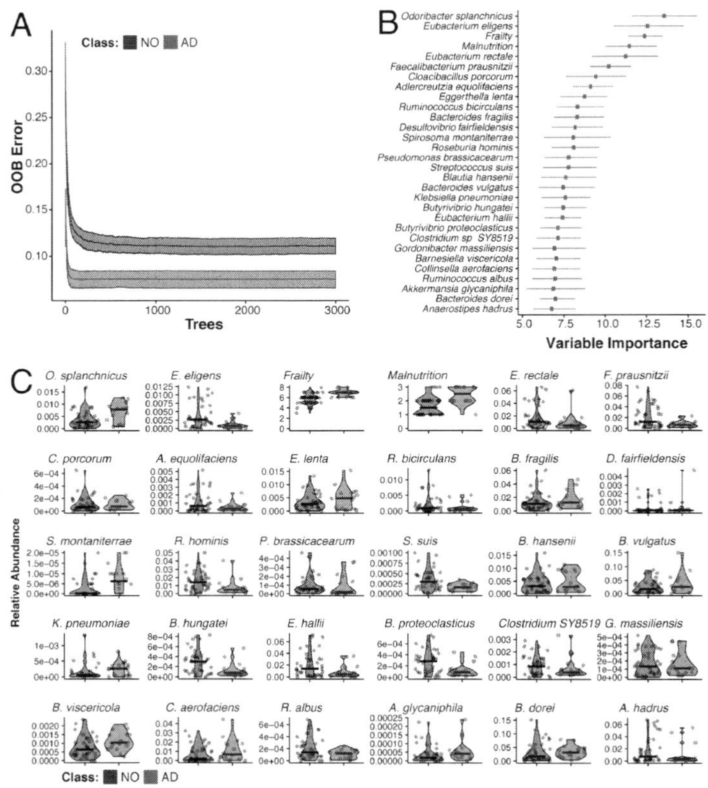

Figure 20: below illustrates how gut microbiota facilitates AD prediction A) AD and NO B) The top-30 discriminating factors C) Abundance of species: Lack of dementia are shown in blue vs. AD (red)(2019 Haran et al: *This is an open-access article distributed under the terms of the Creative Commons Attribution 4.0 International license: You are free to Share — copy and redistribute the material in any medium or format, Adapt — remix, transform, and build upon the material for any purpose, even commercially*).

Parkinson's disease (PD) targets close to 1.5 to 2.0% of the elderly (60 years and above) and ranks as the second most common neurodegenerative disorder. Current treatment aims at controlling the symptoms and augmentation of the quality of life. This warrants further research to explore approaches to address the condition (Marino et al, 2020).

A 2019-published research article looked at the gut microbiota in PD patients vs. that of healthy controls. The major phyla of the gut viz. Firmicutes, Bacteroidetes, Proteobacteria, and Actinobacteria accounted for 98.4% in the control vs. 88.1% in PD patients. As the severity of PD progressed, there was an increase in Gammaproteobacteria (produce lipopolysaccharide [LPS]) and Verrucomicrobiae (that degrade mucin) while Clostridia displayed a declining trend. The patients with mild PD showed an increase of 621.22× fold in Gammaproteobacteria as opposed to controls while the increase of Verrucomicrobiae in severe PD was 4.35× fold against the mild PD cases. The decrease in Clostridia is also important given that these organisms are butyrate-synthesizing commensals and also quell LPS-induced inflammation. The lowering in commensals and increase in Verrucomicrobiae is hypothesized to allow LPS to enter circulation or the enteric nervous system to cause inflammation and aggregation of enteric αSyn. This αSyn or alpha-synuclein aggregation in neurons is archetypical of PD. Interestingly, αSyn increased intestinal permeability and associated inflammatory manifestations.

Thus, the involvement of the gut microbiome along with other factors such as toxins, genetics, drugs, and nutrition to facilitate the αSyn-mediated pathogenesis in PD opens the avenue to comprehend new aspects of these neurodegenerative diseases (Gorecki et al, 2019).

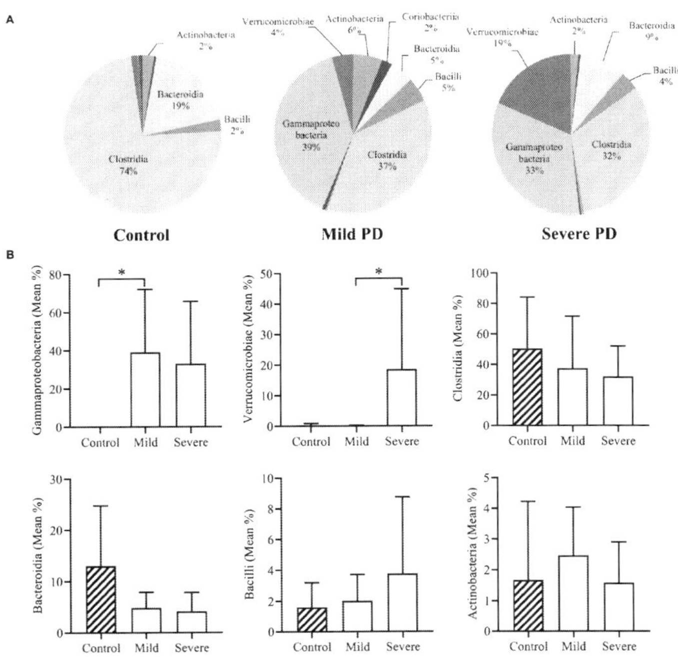

Figure 21: below shows (A) the changes in gut microbiota in PD patients vs. controls and (B) the changes in Gammaproteobacteria and Verrucomicrobiae in mild and severe PD, respectively (2019, Gorecki, Preskey, Bakeberg, Kenna, Gildenhuys, MacDougall, Dunlop, Mastaglia, Akkari, Koengten and Anderton: *This is an open-access article distributed under the terms of the Creative*

Research is uncovering new details of how the microbiota changes in neurodegenerative diseases.

Table 4: below summarizes a few studies and alterations (Suganya K and Koo, 2020: Licensee MDPI, Basel, Switzerland: *This article is an open-access article distributed under the terms and conditions of the Creative Commons Attribution (CC BY) license (http://creativecommons.org/licenses/by/4.0/: You are free to Share — copy and redistribute the material in any medium or format; Adapt — remix, transform, and build upon the material for any purpose, even commercially*)

Neurological Disorder	Models	Changes of Microbiota/Composition	Immune/Inflammatory Effects
AD (Alzheimer's disease)	APP/ PS1 and Tg2576 animals	*Firmicutes, Verrucomicrobiae, Proteobacteria, Actinobacteria, Rikenellaceae* and GS247, *Turicibacteriaceae, Erysipelotrichaceae,* and *Allobaculum* and *Akkermansia Lactobacillus* and *Ruminiclostridium*	Increase in cerebral Aβ pathology and IL-1β production. Plasma MCP-1 was high in symptomatic mice, while IL-9, VEGF-α, and IP-10 were higher in pre-symptomatic Tg2576 mice.
	Human Patients	*Odoribactersplanchnicus, Bacteroides vulgatus, B. fragilis, Eggerthella lenta, Escherichia/Shigella, Gemella, Butyrivibrio, Eubacterium, Clostridium, Roseburia hominis, Bifidobacterium, F. prausnitzii*	Increase in cerebral Aβ accumulation and neuroinflammation. Bacterial LPS was found in the brain. The higher levels of IL-1β, NLRP3, and CXCL2 were positively correlated with *Escherichia/Shigella* abundance.
PD (Parkinson's disease)	Mice	*Verrucomicrobiae, Gammaproteobacteria, Erysipelotrichaceae, Akkermansia, Proteobacteria* and *Lachnospiraceae, Clostridiates*	Microglial activation, inflammatory cytokines, and HO-1 were found to be higher in PD mice. Gut inflammation; disrupted intestinal barrier with a higher level of IL-17, TNF-α, and IL-1β; and activation of microglia, astrocytes, and higher level of TLR4 expression were also found in the brain of PD mice.
	Patients	*Enterobacteriaceae, Proteobacteria, Verrucomicrobiaceae, Lactobacillaceae Porphyromonas, Parabacteroides Prevotellaceae, F. prausnitzii, Bacteroides fragilis, B. dorei, B. pebeus, Ruminococcus callidus*	TLR4, CD3+T cells, cytokines were found to be higher in the intestine of PD patients. Higher levels of IFN-γ, TNF-α were also found in PD patients.

ASD (Autism spectrum disorder)	Patients (Childrens)	*Lactobacillaceae, Bifidobacteraceae, Veillonellaceae, Acidobacteria, Clostridium Acidaminococcaceae, Lachnoclostridium, Flavonifractor, Lacnospiracease, Eubacterium, Rumnicoccaceae, Prevotella copri, F. prausnitzii, H. parainfluenza*	Lower level of fecal acetic acid and butyrate, and higher level of valeric acid were found in ASD subjects. An increase in the level of intestinal serotonin and decrease in level of cerebral serotonin were also found in ASD subjects.
MS (Multiple sclerosis)	Patients	*Desulfovibrionaceae, Haemophilus Verrucomicrobia, Blautia, Dorea, Pseudomonas, Mycoplana, Acinetobacter Enterobacteriaceae, Bacteroidetes, Ruminococcaceae, Heliobcateraceae, Sutterlla Lachnospiraceae, Collinsella*	Serum level of Lipid 654 was lower in MS patients. The immune markers such as Th2, Th17, and Treg did not significantly differed between both controls and MS patients.
Depression	Patients	*Anaerostipes, Blautia, Clostridium, Klebsiella, Lachnospiraceae incertae sedis, Parabacteroides, Parasutterella, Phascolarcobacterium and Streptococcus, Bifidobacterium, Dialister, Escherichia/Shigella, Faecalibacterium, Ruminococcus*	The lower level of neurotransmitter dopamine (DOPAC) and homovanillic acid were found in the depressive patients. In addition, decreased hippocampus 5-HT, BDNF expression and circulatory IL-10, and increased plasma stress hormone were found with depression.

3.10 STUDYING THE CROSS-TALK: THE WAY AHEAD

Currently, the stage is set for further research given the findings of how the gut microbiota and the neural systems interact with each other. In order to delve into the details further, several approaches have been are explored. The details or changes in bacterial species of the gut can be assessed with 16s rRNA gene sequence analyses and/or high-throughput sequencing. Analytical approaches entailing biochemistry can permit the identification of specific molecules or mediators. Further, detailed analyses of the gut-brain microbiome axis can be facilitated by advanced neuroimaging methods. This can encompass structural MRI to functional MRI (fMRI), magnetic resonance spectroscopy (MRS), diffusion tensor imaging (DTI), and mapping iron in the brain.

Table 5: below lists few studies that used these various imaging approaches to gain insights on the gut-brain axis (2019 Liu, Peng, Zhang, Wang and Luo.

Subjects	Intervention	Measures	Results				References
			Neuroimaging results	GM results	Correlation	Other	
Healthy women	FMPP for 4w (FMPP group, *n* = 12; non-FMPP group, *n* = 12; no intervention, *n* = 13)	Task-based fMRI and rs-fMRI GM (fecal samples)	Emotional attention task based-fMRI: sensory brain network connection strength and decreases in insular and somatosensory cortical BOLD activity↓ in the FMPP group rs-fMRI: FAG was negatively correlated with sensory/affective regions and positively correlated with cortical regulatory regions (medial and dorsolateral prefrontal cortices) in FMPP group	No significant changes in fecal microbiota composition	Four-week intake of an FMPP by healthy women affected activity in brain regions that control the central processing of emotion and sensation	/	(25)
Obese and non-obese subjects	No intervention (20 obese and 19 non-obese subjects)	MRI: DTI; FLAIR; R2*; GM (fecal samples); cognitive tests	See the correlation column	16S bacterial gene pyrosequencing: fecal sample bacterial biodiversity↓ in obese men	Fecal microbiota diversity was negatively correlated with R2* signals in the hypothalamus, hippocampus, and caudate nucleus The abundance of Actinobacteria was positively associated with FA in the amygdala and thalamus but negatively correlated with the R2* signal in the hypothalamus	The relative abundance of the Actinobacteria Phylum was positively associated with cognitive tests related to speed, attention, and cognitive flexibility	(28)
Elderly outpatients with and without cirrhosis	No intervention Group type 1: 39 cirrhotic and 37 non-cirrhotic patients. Group type 2: unimpaired cognition (*n* = 23), amnestic-type (*n* = 25), and amnestic/non-amnestic type (*n* = 28).	Multi modal MRI (fMRI go/no-go task, volumetry, and MRS); inflammatory cytokines; GM (fecal samples); neuropsychological tests	No significant fMRI differences in brain volumes between cirrhotic and non-cirrhotic subjects Amnestic/non-amnestic type activation in the central opercular cortex, post-central gyrus, and superior parietal lobule during inhibition↑. Amnestic-type type: white matter, grey matter, and total brain volumes↓, hippocampal and left thalamic volumes↓ Cirrhotic subjects: mI/Cr and NAA/Cr ratios↓ and Glx/Cr ratio↑ Amnestic/non-amnestic type: mI/Cr and Glx/Cr ratios↓	Cirrhotic subjects: *Lactobacillales*↑ and *Synergistaceae*, and *Peptococcaceae*↓. Cognitively impaired groups: decreased *Subdoligranulum*, *Oscillibacter*, and *Porphyromonadaceae* and *Prevotellaceae*↓ and *Bacteroides*↑ Unimpaired group: increased *Fecalibacterium* and *Butyricicoccus*	Regardless of the presence of cirrhosis, beneficial taxa (*Lactobacillales*, *Ruminococcaceae*, and *Lachnospiraceae*) were positively linked with cognition while pathogenic taxa (*Enterobacteriaceae*) were negatively linked with cognition	Serum levels of IL-6/endotoxin↑	(37)
Cirrhosis patients with/without prior HE	No intervention (Cirrhotic without prior HE, *n* = 62; cirrhotic with prior HE, *n* = 85; controls, *n* = 40)	MRS, DTI, Systemic Inflammatory assessment, GM (fecal samples); cognitive testing	MRS: Cirrhotic patients with HE: Glx↑↑ mI↓↓, Cho↓↓. Cirrhotics: Glx↑ mI↓; DTI: Cirrhotics with HE: spherical isotropy↑, FA↓	Cirrhotic patients with HE *Staphylococcaceae*↑ *Enterococcaceae*↑, *Porphyromonadaceae*↑, and *Lactobacillaceae*↑; autochthonous bacterial families (*Lachnospiraceae*↓, *Ruminococcaceae*↓, and *Clostridiales XIV*↓	Autochthonous taxa were negatively correlated with hyperammonemia associated astrocyte MRS changes while *Enterobacteriaceae* were positively correlated with hyperammonemia-associated astrocytic MRS changes (high Glx levels and low mI and Cho levels). *Porphyromonadaceae* were correlated with neuronal changes on DTI without being linked to ammonia	Cirrhotic patients with prior HE had significantly more advanced cirrhosis, more severe and higher levels of inflammatory markers and cognitive impairments compared to cirrhotic patients without HE	(29)
IBS patients with anxiety and depression	Probiotic: *Bifidobacterium longum* NCC3001 for 6 w (BL, *n* = 22; placebo, *n* = 22)	Task based fMRI (fearful face backward masking paradigm); fecal microbiota; urine metabolome profiles, serum inflammatory markers; neurotransmitter and neurotrophin levels	BL reduced responses to negative emotional stimuli in multiple brain areas including the amygdala and fronto-limbic regions	No major changes in fecal microbiota composition after intervention	Reduced amygdala activity was correlated with decreases in depression scores	Depression scores↓; IBS symptoms improved; urine levels of methylamines and aromatic amino acid metabolites↓	(49)
Healthy women (*Bacteroides*-high group vs. *Prevotella*-high group)	No intervention	Emotion-induced task-based fMRI; structural MRI (DTI, T1)	*Prevotella* group: right hippocampal activity↓ when viewing negative valence images. *Bacteroides* high group: white matter connectivity↓, cerebellum, frontal region, and hippocampal volumes↑, and nucleus accumbal volume↓	Subjects were divided into *Bacteroides*-high group and *Prevotella*-high group based on fecal microbiota analysis	/		(50)
Obese and non-obese subjects	Diet counseling (18 obese and 17 non-obese subjects)	MR relaxometry R2*; GM (fecal samples); Neuropsychological tests: plasma β-amyloid (1–24, 31–40) levels	MR R2* relaxometry increased mainly in the pallidum, putamen, thalamus, and hippocampus in both groups over a 2-year period	A variety of gut microbiome changes in RA over a 2-year period	Shifts in *Gemmatimonadetes*, *Bacteroidetes*, *Proteobacteria*, *Cyanobacteria*, *Candidatus*, *Saccharibacteria*, *Tenericutes*, *Thermodesulfobacteria*, and *Chlorobi* RA were associated with increased percentages of R2* in the striatum, superficial amygdala, and hippocampus. Shifts in the phyla *Fibrobacteres*, *Synergistetes*, and *Tenericutes* RA were reciprocally associated with right hippocampal R2*.	Circulating β-amyloid Ab42 levels were positively associated with changes in visuospatial constructional ability and immediate memory but negatively associated with increases in R2*	(43)
Healthy volunteers	Probiotic (Ecologic825 nine bacterial strains) for 4w (probiotic, *n* = 15; placebo, *n* = 15; no intervention, *n* = 15)	Task-based fMRI (emotional decision-making and emotional recognition memory); DTI; neuropsychological tests	Altered brain activation in the cingulum, precuneus, inferior parietal lobule, thalamus, and parahippocampal gyrus in the ED task and cerebellar activity in the ER task	No major changes in general fecal microbial diversity or evenness; *Bacteroides sp.*↑, *Alistipe sp.*↑, and the nicotinate and nicotinamide metabolic pathway of fecal microbiota↓	Probiotic ingestion improved emotional attention and memory performance, which was accompanied by changes in activity in corresponding brain regions	Self-reported behavioral measures of positive affect, cognitive reactivity, and memory performance improved	(59)
Healthy volunteers	Probiotic (Ecologic825 nine bacterial strains) for 4w (probiotic, *n* = 15; placebo, *n* = 15; no intervention, *n* = 15)	Rs-fMRI; diffusion MRI	FC in MFGN (in frontal pole and frontal medial cortex) and in DMN (in frontal lobe)↓, FC in SN (in cingulate gyrus and precuneus cortex)↑. No significant changes in structural connectivity (FA/MD)	Same as above (Bagga et al. (59))	Probiotic intervention was associated with changes in FC but not structural connectivity		(51)
High-risk (HR) and ultra-high risk (UHR) subjects for schizophrenia	No intervention (high-risk group, *n* = 81; ultra-high risk group, *n* = 19; healthy controls, *n* = 69)	MRS; GM (fecal samples).	Ultra-high risk group: Cho levels in anterior cingulate cortex↑	Ultra-high risk group: at order level; *Clostridiales*, *Lactobacillales*, and *Bacteroidales*↑; at general level, *Prevotella* and *Lactobacillus*↑, synthesis of acetyl-CoA (belonging to SCFAs)↑	Alterations in MRS and GM function (synthesis of SCFAs) support the hypothesis that membrane dysfunction exists in schizophrenia	/	(52)

Summary of multimodal MR imaging studies investigating the relationship between the adult GM and brain function

Subjects	Intervention	Measures	Results				References
			Neuroimaging results	GM results	Correlation	Other	
Healthy women	FMPP for 4w (FMPP group, $n = 12$; non-FMPP group, $n = 12$; no intervention, $n = 13$)	Task-based fMRI and rs-fMRI; GM (fecal samples)	Emotional attention task based-fMRI: sensory brain network connection strength and decreases in insular and somatosensory cortical BOLD activity↓ in the FMPP group rs-fMRI: PAG was negatively correlated with sensory/affective regions and positively correlated with cortical regulatory regions (medial and dorsolateral prefrontal cortices) in FMPP group	No significant changes in fecal microbiota composition	Four-week intake of an FMPP by healthy women affected activity in brain regions that control the central processing of emotion and sensation	/	(25)
Obese and non obese subjects	No intervention (20 obese and 19 non obese subjects)	MRI; DTI; FLAIR; R2*; GM (fecal samples); cognitive tests	See the correlation column	16S bacterial gene pyrosequencing: fecal sample bacterial biodiversity↓ in obese men	Fecal microbiota diversity was negatively correlated with R2* signals in the hypothalamus, hippocampus, and caudate nucleus. The abundance of Actinobacteria was positively associated with FA in the amygdala and thalamus but negatively correlated with the R2* signal in the hypothalamus	The relative abundance of the Actinobacteria Phylum was positively associated with cognitive tests related to speed, attention, and cognitive flexibility	(26)
Elderly outpatients with and without cirrhosis	No intervention Group type 1: 39 cirrhotic and 37 non-cirrhotic patients; Group type 2: unimpaired cognition ($n = 23$), amnestic-type ($n = 25$), and amnestic/non amnestic type ($n = 28$).	Multi-modal MRI (fMRI go/no-go task, volumetry, and MRS); inflammatory cytokines; GM (fecal samples); neuropsychological tests	No significant fMRI differences in brain volumes between cirrhotic and non-cirrhotic subjects. Amnestic/non-amnestic type: activation in the central opercular cortex, post-central gyrus, and superior parietal lobule during inhibition↑. Amnestic-type type: white matter, gray matter, and total brain volumes↓, hippocampal and left thalamic volumes↓ Cirrhotic subjects: mI/Cr and NAA/Cr ratios↓ and Glx/Cr ratio↑ Amnestic/non-amnestic type: mI/Cr and Glx/Cr ratios↓	Cirrhotic subjects: *Lactobacillales*↑ and *Synergisticeae*, and *Peptococcaceae*↓. Cognitively impaired groups: decreased *Subdoligranulum*, *Oscillibacte*, and *Porphyromonadaceae* and *Prevotellaceae*↓ and *Bacteroides*↑ Unimpaired group: increased *Fecalibacterium* and *Butyricicoccus*	Regardless of the presence of cirrhosis, beneficial taxa (*Lactobacillales*, *Ruminococcaceae*, and *Lachnospiraceae*) were positively linked with cognition while pathogenic taxa (*Enterobacteriaceae*) were negatively linked with cognition	Serum levels of IL-6/endotoxin↑	(27)
Cirrhosis patients with/without prior HE	No intervention (Cirrhotic without prior HE, $n = 62$; cirrhotic with prior HE, $n = 85$; controls, $n = 40$)	MRS, DTI, Systemic inflammatory assessment, GM (fecal samples); cognitive testing	MRS: Cirrhotic patients with HE: Glx↑↑ mI↓↓; Cho↓↓; Cirrhotics: Glx↑ mI↓; DTI: Cirrhotics with HE: spherical isotropy↑, FA↓	Cirrhotic patients with HE *Staphylococcaceae*↑, *Enterococcaceae*↑, *Porphyromonadaceae*↑, and *Lactobacillaceae*↑, autochthonous bacterial families (*Lachnospiraceae*↓, *Ruminococcaceae*↓, and *Clostridiales XIV*↓	Autochthonous taxa were negatively correlated with hyperammonemia-associated astrocytic MRS changes while *Enterobacteriaceae* were positively correlated with hyperammonemia-associated astrocytic MRS changes (high Glx levels and low mI and Cho levels). *Porphyromonadaceae* were correlated with neuronal changes on DTI without being linked to ammonia	Cirrhotic patients with prior HE had significantly more advanced cirrhosis, more severe and higher levels of inflammatory markers and cognitive impairments compared to cirrhotic patients without HE	(28)
IBS patients with anxiety and depression	Probiotic: *Bifidobacterium longum* NCC3001 for 6 w (BL, $n = 22$, placebo, $n = 22$)	Task-based fMRI (fearful face backward masking paradigm); fecal microbiota; urine metabolome profiles; serum inflammatory markers; neurotransmitter and neurotrophin levels	BL reduced responses to negative emotional stimuli in multiple brain areas including the amygdala and fronto-limbic regions	No major changes in fecal microbiota composition after intervention	Reduced amygdala activity was correlated with decreases in depression scores	Depression scores↓; IBS symptoms improved; urine levels of methylamines and aromatic amino acid metabolites↓	(29)
Healthy women (*Bacteroides*-high group vs. *Prevotella*-high group)	No intervention	Emotion-induced task-based fMRI; structural MRI (DTI, T1)	*Prevotella* group: right hippocampal activity↓ when viewing negative valence images. *Bacteroides*-high group: white matter connectivity↓, cerebellum, frontal region, and hippocampal volumes↑, and nucleus accumbal volume↓	Subjects were divided into *Bacteroides*-high group and *Prevotella*-high group based on fecal microbiota analysis	/		(30)
Obese and non-obese subjects	Diet counseling (18 obese and 17 non-obese subjects)	MR relaxometry R2*; GM (fecal samples); Neuropsychological tests; plasma β-amyloid (1-24, 31-42) levels	MR R2* relaxometry increased mainly in the pallidum, putamen, thalamus, and hippocampus in both groups over a 2-year period	A variety of gut microbiome changes in RA over a 2-year period	Shifts in *Gemmatimonadetes*, *Bacteroidetes*, *Proteobacteria*, *Caldiserica*, *Candidatus*, *Saccharibacteria*, *Tenericutes*, *Thermodesulfobacteria*, and *Chlorobi* RA were associated with increased percentages of R2* in the striatum, superficial amygdala, and hippocampus. Shifts in the phyla *Fibrobacteres*, *Synergistetes*, and *Tenericutes* RA were reciprocally associated with right hippocampal R2*.	Circulating β-amyloid Ab42 levels were positively associated with changes in visuospatial constructional ability and immediate memory but negatively associated with increases in R2*	(49)
Healthy volunteers	Probiotic (Ecologic®825, nine bacterial strains) for 4w (probiotic, $n = 15$; placebo, $n = 15$; no intervention, $n = 15$)	Task-based fMRI (emotional decision making and emotional recognition memory); DTI; neuropsychological tests	Altered brain activation in the cingulum, precuneus, inferior parietal lobule, thalamus, and parahippocampal gyrus in the ED task and cerebellar activity in the ER task	No major changes in general fecal microbial diversity or evenness: *Bacteroides sp.*↑, *Alistipe sp.*↑, and the nicotinate and nicotinamide metabolic pathway of fecal microbiota↓	Probiotic ingestion improved emotional attention and memory performance, which was accompanied by changes in activity in corresponding brain regions	Self-reported behavioral measures of positive affect, cognitive reactivity, and memory performance improved	(50)
Healthy volunteers	Probiotic (Ecologic®825, nine bacterial strains) for 4w (probiotic, $n = 15$; placebo, $n = 15$; no intervention, $n = 15$)	Rs-fMRI; diffusion MRI	FC in MFGN (in frontal pole and frontal medial cortex) and in DMN (in frontal lobe)↓, FC in SN (in cingulate gyrus and precuneus cortex)↑. No significant changes in structural connectivity (FA/MD)	Same as above (Bagga et al. (50))	Probiotic intervention was associated with changes in FC but not structural connectivity		(51)
High-risk (HR) and ultra-high risk (UHR) subjects for schizophrenia	No intervention (high-risk group, $n = 81$; ultra-high risk group, $n = 19$; healthy controls, $n = 69$)	MRS; GM (fecal samples).	Ultra-high risk group: Cho levels in anterior cingulate cortex↑	Ultra-high risk group at order level: *Clostridiales*, *Lactobacillales*, and *Bacteroidales*↑; at general level, *Prevotella* and *Lactobacillus*↑; synthesis of acetyl-CoA (belonging to SCFAs)↑	Alterations in MRS and GM function (synthesis of SCFAs) support the hypothesis that membrane dysfunction exists in schizophrenia	/	(52)

Summary of multimodal MR imaging studies investigating the relationship between the adult GM and brain function

Interestingly, the concept of "radiomicrobiomics" has been presented. This entails the diagnosis of diseases employing microbiome sequencing and brain imaging data (De santis et al, 2019). If we look at iron metabolism and nervous system disease, Lin and team (2013) assessed the level of hepcidin in patients with hepatitis B virus (HBV)-related cirrhotic patients. The molecule hepcidin produced by the liver is vitally involved in iron homeostasis. Given the neurological complications associated with hepatic cirrhosis, the level of iron in the brain of these patients was assessed by susceptibility weighted (SW) phase imaging (an MRI approach). They found that the brain of cirrhosis patients displayed elevated levels of iron as opposed to healthy controls in the left red nucleus, bilateral substantia nigra, the bilateral thalamus, the right caudate, and the right putamen. Further, the hepcidin levels were lower in these patients suggestive of disequilibrium in iron metabolism. This becomes interesting as *Bacteroides fragilis* and *Bifidobacterium longum*, our gut microbiota members can influence the expression of hepacidin (Liu et al, 2019).

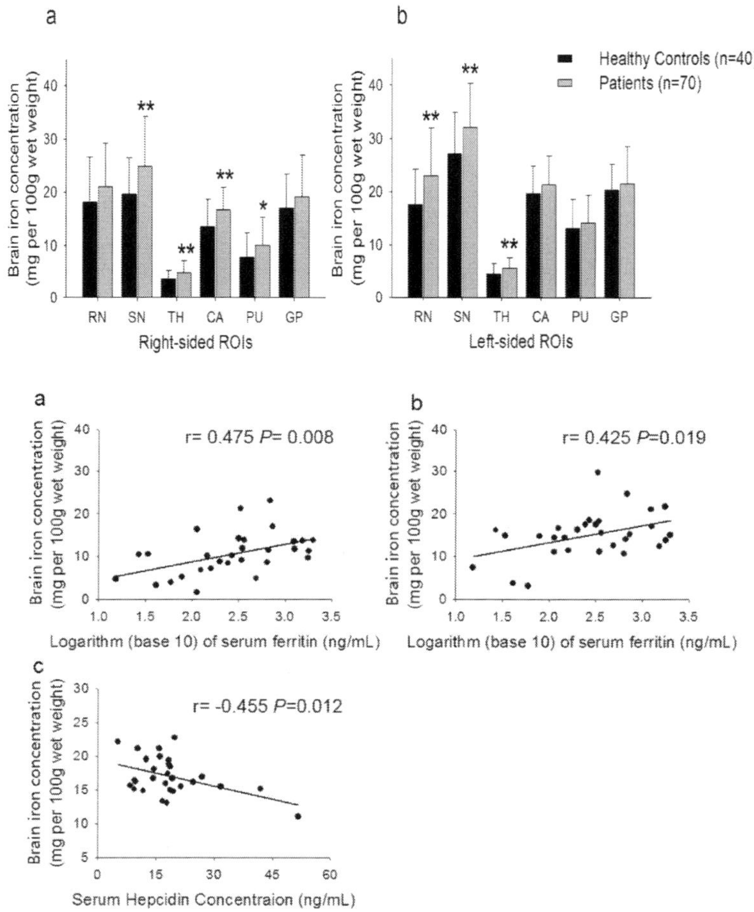

Figure 22: Above shows ***TOP***: Iron levels in the brains of cirrhosis patients and controls (a) left (b) right hemisphere; ***BOTTOM***: Negative correlation between serum hepcidin level and the iron content in the right caudate while a positively correlation was seen in the bilateral putamen in cirrhotic patients (2013 Lin et al: *This is an open-access article distributed under the terms of the Creative Commons Attribution License, which permits unrestricted use, distribution, and*

reproduction in any medium, provided the original author and source are credited).

The cross-talk between the gut microbes and the brain is just being unraveled with new research possibly uncovering the various details. An amalgamation of biochemical, imaging and OMICS approaches can allow us to comprehend the connection between the gut and the brain.

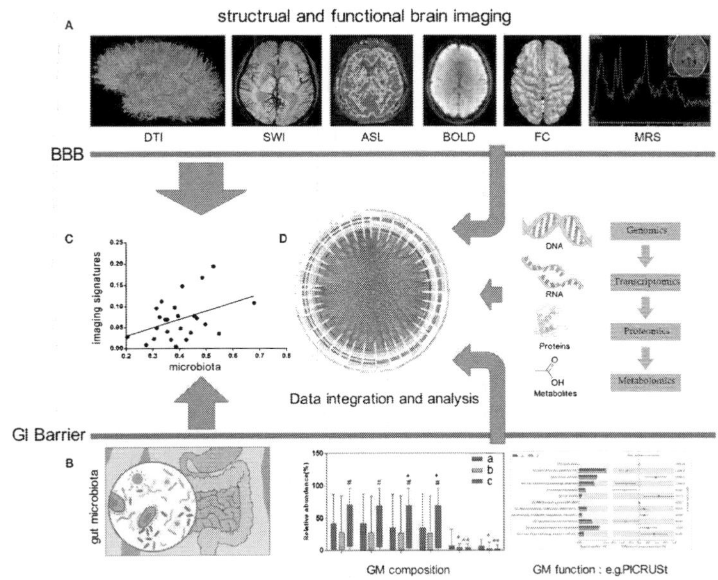

Figure 23: Above illustrates the way ahead to explore how gut bacteria influence the nervous system. (A) MRI maps (B) Analyzing the microbiota of the gut (C) Imaging-based studies (D) Including OMICS*; DTI, Diffusion Tensor Imaging; SWI, Susceptibility Weighted Imaging; ASL, Arterial Spin Labeling; BOLD, blood oxygenation level dependent; FC, functional connection; MRS, Magnetic Resonance Spectroscopy; BBB, blood-brain barrier; GI, gastro-intestinal; GM, gut microbiota; PICRUSt, Phylogenetic Investigation of Communities by Reconstruction of Unobserved States* (2019 Liu, Peng, Zhang, Wang and Luo. *This is an open-access article distributed under the terms of the* ***Creative Commons Attribution License (CC BY)****. The use, distribution or reproduction in other forums is permitted, provided the original author(s) and the copyright owner(s) are credited and that the original publication in this journal is cited, in accordance with accepted academic practice. No use, distribution or reproduction is permitted which does not comply with these terms*).

There are several metabolites that are being uncovered by research to possibly pass across the blood-brain barrier to influence the CNS.

Table 6: below illustrates few such metabolites identified (2020 Raimondi, Izzo, Tunesi, Comar, Albani and Giordano.

Neuro-modulators	Bacterial genera	Brain disease	References
5-Hydroxytryptamine (5-TH)	*Candida, Enterococcus, Escherichia, Streptococcus*	anxiety, depression, MS, PD	Baganz and Blakely, 2013; Mawe and Hoffman, 2013; O'Mahony et al., 2015; Yano et al., 2015
β-Methylamino-L-alanine (BMAA)	*Cyanobacteria*	AD, ALS, PD	Meneely et al., 2016; Delcourt et al., 2017
γ-Aminobutyric (GABA)	*Bifidobacterium, Lactobacillus*	AD, anxiety, depression	Barrett et al., 2012; Lin, 2013; Hu et al., 2016
Acetylcholine	*Lactobacillus*	AD	Wessler and Kirkpatrick, 2008; Wall et al., 2014
Catecholamine (Adrenaline, Dopamine, Noradrenaline)	*Bacillus, Escherichia, Saccharomyces*	ASD, depression, PD, schizophrenia	Mayer and Hsiao, 2017; Mittal et al., 2017; Sugame et al., 2017
Histamine	*Lactobacillus, Lactococcus, Pediococcus, Streptococcus, Enterococcus spp.*	AD, MS	Landete et al., 2008; Thomas et al., 2012; Naddafi and Mirshafiey, 2013; Westfall et al., 2017
Lipopolysaccharide (LPS)	*Gram-negative bacteria*	AD, anxiety, ASD depression, HD, MS, PD, schizophrenia, etc.	Zhao et al., 2017
Short-chain fatty acids (SCFA, e.g., acetate, propionate and butyrate)	*Bacteroides, Bifidobacterium, Clostridium, Eubacterium, Lactobacillus, Propionibacterium, Roseburia Prevotella*	AD, ASD, HD, PD	Liu et al., 2015; Verbeke et al., 2015; Bourassa et al., 2016; Ho et al., 2018; Van de Wouw et al., 2018
Gingipains	*Porphyromonas gingivalis*	AD, PD	Adams et al., 2019; Dominy et al., 2019

Summary of the most important neuropeptides involved in brain disorders such as anxiety, autism, AD, amyotrophic lateral sclerosis, depression, MS, schizophrenia, and PD.

This brings in Organ-on-a-Chip-Based *in vitro* models to understand the processes of the blood-brain barrier. An example is a chip composed of a 10 μm thick Transwell®-like polycarbonate membrane separating two PDMS (polydimethylsiloxane) layers that allows the culture of CMEC/D3 EC (a blood-brain barrier cell line) with Pt electrodes to quantify the TEER (trans-endothelial electrical resistance or trans-epithelial electrical resistance) (Griep et al, 2013).

Organ-on-a-Chip-Based *in vitro* models allow for exposure to stress or flow and allow cell interactions that are spatially-oriented. The small size facilitates lesser use of reagents while allowing electric measuring devices to be integrated for quantifying various parameters (Raimondi et al, 2020).

The MINERVA (ID 724734) project by the European Research Council is aimed at assessing the role of the gut microbiota on neurodegeneration. This microfluidic organs-on-a-chip approach would comprise the microbiota along with the brain and blood-brain barrier. The secretome from the gut microbiota of donors (ill or healthy) cultured on an intestinal mucus-simulating hydrogel matrix would be the first chamber. This is followed by gut epithelial cells on a microporous membrane representative of the gut barrier. This is followed by the immune cells lymphocytes and macrophages with the next as endothelial cells and astrocytes to represent the blood-brain brain and finally a 3D hydrogel modeling brain with neurons, astrocytes and microglia. This *in vitro* multi-organ platform is a future path to explore non-invasive approaches to comprehend the intricacies of the microbiota (Raimondi et al, 2020).

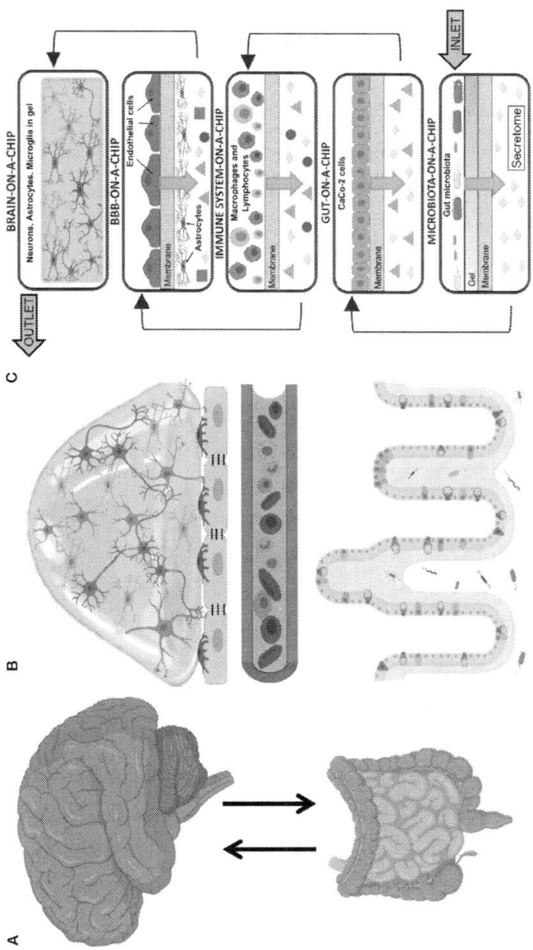

Figure 24: Above a scheme of the various facets of the gut-brain axis and its analyses by current technology such as the organ-on-a-chip platform: MINER-

VA. The details of the cell layers to simulate the various organ systems have been discussed above (2020 Raimondi, Izzo, Tunesi, Comar, Albani and Giordano. *This is an open-access article distributed under the terms of the* ***Creative Commons Attribution License (CC BY)****. The use, distribution or reproduction in other forums is permitted, provided the original author(s) and the copyright owner(s) are credited and that the original publication in this journal is cited, in accordance with accepted academic practice. No use, distribution or reproduction is permitted which does not comply with these terms*).

3.11 FINAL POINTS

The current findings and research on gut microbiota on diseases of the nervous system are summarized below in

Table 7: (Zhu et al, 2020: *Open-Access: This article is distributed under the terms of the Creative Commons Attribution 4.0 International License (http://creativecommons.org/licenses/by/4.0/), which permits unrestricted use, distribution, and reproduction in any medium, provided you give appropriate credit to the original author(s) and the source, provide a link to the Creative Commons license, and indicate if changes were made. The Creative Commons Public Domain Dedication waiver (http://creativecommons.org/publicdomain/zero/1.0/) applies to the data made available in this article, unless otherwise stated*):

Brain disorders	Dysbiosis of microbiota	Pathological and molecular signature	Potential probiotic-based therapeutics
Stress and depression	Increase of *Faecalibacterium, Alistipes, Ruminococcus* [112], *Campylobacter jejuni* [143] and *Firmicutes*; decrease of *Bacteroidetes* [251]	Activation and releasing of HPA [121],VFA [112, 141], serotonin 1A receptor [122]; reduced expression of BDNF [122], GABA receptor [119]	*Bifidobacterium infantis* [121], *Lactobacillus rhamnosus* [119], *Lactobacillus helveticus, Bifidobacterium longum* [142], and other *Bifidobacterium* strains [143]
Pain and migraine	Increase of *H.pylori* [162]; dysbiosis [163]	Dysfunction of CB2 and TLR7 [160]; inflammatory hypernociception [161]; TNF-α induced chronic migrain e[163]	*Lactobacillus, B. infantis* 35624 and *Lactobacillus farciminis* [165, 166], *Lactobacillus salivarius* UCC433 1[167]

Autism spectrum disorders	Decreases of *Faecalibacterium* species [150], *Bifidobacteria* species, *Akkermansia muciniphila*; increase of *Lactobacillus, Bacteroides, Prevotella*, and *Alistipes* [145, 151, 152]; alteration of *Fusobacteria, Verrucomicrobia* and *Firmicutes/Bacteroides* ratio [11]	Disrupted intestinal epithelial permeability [114], mucosal inflammation [148, 149], reduction of SCFA synthesis [150], mucosal inflammation [148, 149]	*Bacteroides fragilis* [114], *Lactobacillus* species and *Bifidobacterium* species [119, 155]
Parkinson's disease	Increase of *H.pylori* [174], *E.coli* [179], *Ralstonia, Oscillospira* and *Bacteroides* [180]; decrease of *Prevotellaceae* [181], *Blautia, Coprococcus*, and *Roseburia* [180]	Higher frequency of α-synuclein detection [175], dopaminergic neuronal loss [177], bowel inflammation and change of permeability [178, 180]	Multiple probiotics strains (*Streptococcus salivarius subsp thermophilus, Enterococcus faecium, Lactobacillus rhamnosus GG, Lactobacillus acidophilus, Lactobacillus plantarum, Lactobacillus paracasei, Lactobacillus delbrueckii subsp bulgaricus*, and *Bifidobacterium*) [182]
Alzheimer's disease	Chronic *H.pylori* infection [193, 194]; increase of *E.coli, Salmonella* spp, *Pseudomonas fluorescens, Klebsiella pneumonia, Staphylococcus aureus, Bacillus subtills, Streptomyces coelicolor* [190, 196, 197]; *Chlamydia pneumoniae* infection [199]	Increased levels of IL-1beta, IL-6, and TNF-alpha [186, 187]; extracellular amyloids such as CsgA, Aβ42, FapC, MccE492, phenol-soluble Modulins, TasA, Chaplins, and β-amyloid exhibit cerebral deposition [190, 192, 196, 197, 199]	*Lactobacilli* and *Bifidobacteria* [200, 203]; *Lactobacillus fermentum NS9* and *Lactobacillus helveticus NS8* [201, 202]
Amyotrophic lateral sclerosis	Reduced levels of butyrate-producing bacteria, including *Butyrivibrio fibrisolvens, Escherichia coli, Oscillibacter, Anaerostipes*, and *Lachnospira* [213]; increase of glucose-metabolizing *Dorea* [214]	Tetanus and botulinum toxins [210, 211]; "leaky gut" [212]; higher richness of OTUs [215]; reduction of butyrate [216]	*Lactobacillus* strains [217]
Multiple sclerosis	Increase of *Archaea, Psuedomonas, Haemophilus, Blautia, Dorea* [219], and *Fusobacteria* [225]; decrease of *Bacteroidetes phyla, Firmicutes phyla,* [218] *Parabacteroides, Adlercreutzia, Prevotella* [219], *Bacteroides* and *Clostridia* [218, 222]	BBB integrity disruption and astrocyte pathogenicity [222]; increasing proinflammatory cell infiltration and impaired Treg function [221]; "leaky gut" [219, 220]	*Bifidobacterium* and lactic acid-producing bacteria [223, 224]; *Bacteroidaceae, Faecalibacterium, Ruminococcus, Lactobacillaceae*, and *Clostridium* [226]

Atherosclerosis	*Lactobacillus rhamnosus* and *Neisseria polysaccharea, Acidovorax* spp. and *H. pylori* cells [232]; *Collinsella* [233]; *Roseburia* and *Eubacterium* [233]	Carnitine-butyrobetaine-trimethylamine-*N*-oxide pathway [235]; cause vascular atherosclerosis by affecting lipid and hormonal homeostasis [82, 222]; formation of foam cell by inflammation [236, 237]	*L. acidophilus* 145, *B. longum* 91 3[238], *L. acidophilus* and *B. bifidum* [239], *L. plantarum ZDY04* against TMAO [240]
Stroke	Increase of *Porphyromonas gingivalis* [242], Gram-negative bacteria [243], *Enterobacter, Megasphaera, Oscillibacter*; decrease of *Bacteroides, Prevotella* and *Faecalibacterium* [245]	Formation of blood clots and platelet aggregation [244]; decrease of regulatory T cells [241]	Antibiotic administration [241]
Arteriovenous malformation	Gram-negative bacteria [248]	Macrophage infiltration and cytokine secretion [246]; activate TLR4 [248]	Antibiotic and probiotics treatment are controversial [248,249,250]

Associations of a lucid kind are being unveiled by research between our physiology and the gut microbiota. These organisms control both the intestine and the brain. The processes of neurophysiology and cognition are controlled by the gut microbes to influence neurotransmission via their produced metabolites, immune cells and hormones.

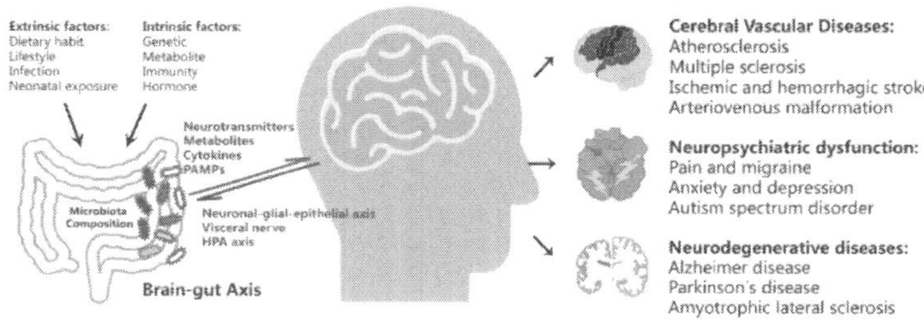

Figure 25: Above is illustrative of the internal and extrinsic factors to alter the gut microbes and result in CNS abnormalities (Zhu et al, 2020: *Open-Access: This article is distributed under the terms of the Creative Commons Attribution 4.0 International License (http://creativecommons.org/licenses/by/4.0/), which permits unrestricted use, distribution, and reproduction in any medium, provided you give appropriate credit to the original author(s) and the source, provide a link to the Creative Commons license, and indicate if changes were*

made. The Creative Commons Public Domain Dedication waiver (http://creativecommons.org/publicdomain/zero/1.0/) applies to the data made available in this article, unless otherwise stated).

This resonates with the following quote:

"The road to health is paved with good intestines!"

— Sherry A. Rogers

3.12 REFERENCES:

1. Ma, Q., Xing, C., Long, W. *et al.* Impact of microbiota on central nervous system and neurological diseases: the gut-brain axis. *J Neuroinflammation* 16, 53 (2019). https://doi.org/10.1186/s12974-019-1434-3
2. Suganya K, Koo B-S. Gut–Brain Axis: Role of Gut Microbiota on Neurological Disorders and How Probiotics/Prebiotics Beneficially Modulate Microbial and Immune Pathways to Improve Brain Functions. *International Journal of Molecular Sciences*. 2020; 21(20):7551. https://doi.org/10.3390/ijms21207551
3. Lu L, Yu Y, Guo Y, Wang Y, Chang EB, Claud EC (2015) Transcriptional Modulation of Intestinal Innate Defense/Inflammation Genes by Preterm Infant Microbiota in a Humanized Gnotobiotic Mouse Model. PLoS ONE 10(4): e0124504. https://doi.org/10.1371/journal.pone.0124504
4. Lu, J., Lu, L., Yu, Y. *et al.* Effects of Intestinal Microbiota on Brain Development in Humanized Gnotobiotic Mice. *Sci Rep* 8, 5443 (2018). https://doi.org/10.1038/s41598-018-23692-w
5. Wang Y, Wang Z, Wang Y, Li F, Jia J, Song X, Qin S, Wang R, Jin F, Kitazato K and Wang Y (2018) The Gut-Microglia Connection: Implications for Central Nervous System Diseases. *Front. Immunol.* 9:2325. doi: 10.3389/fimmu.2018.02325
6. David, L. A., Maurice, C. F., Carmody, R. N., Gootenberg, D. B., Button, J. E., Wolfe, B. E., Ling, A. V., Devlin, A. S., Varma, Y., Fischbach, M. A., Biddinger, S. B., Dutton, R. J., & Turnbaugh, P. J. (2014). Diet rapidly and reproducibly alters the human gut microbiome. *Nature*, *505*(7484), 559–563. https://doi.org/10.1038/nature12820
7. Ong, I. M., Gonzalez, J. G., McIlwain, S. J., Sawin, E. A., Schoen, A. J., Adluru, N., Alexander, A. L., & Yu, J. J. (2018). Gut microbiome populations are associated with structure-specific changes in white matter architecture. *Translational psychiatry*, *8*(1), 6. https://doi.org/10.1038/s41398-017-0022-5
8. Li L, Su Q, Xie B, Duan L, Zhao W, Hu D, Wu R, Liu H. Gut microbes in correlation with mood: case study in a closed experimental human life support system. Neurogastroenterol Motil. 2016 Aug;28(8):1233-40. doi: 10.1111/nmo.12822. Epub 2016 Mar 29. PMID: 27027909.

9. Park, H. R., Lee, J. M., Moon, H. E., Lee, D. S., Kim, B. N., Kim, J., Kim, D. G., & Paek, S. H. (2016). A Short Review on the Current Understanding of Autism Spectrum Disorders. *Experimental neurobiology*, *25*(1), 1–13. https://doi.org/10.5607/en.2016.25.1.1
10. De Angelis M, Piccolo M, Vannini L, Siragusa S, De Giacomo A, Serrazzanetti DI, et al. (2013) Fecal Microbiota and Metabolome of Children with Autism and Pervasive Developmental Disorder Not Otherwise Specified. PLoS ONE 8(10): e76993. https://doi.org/10.1371/journal.pone.0076993
11. Kang, D. W., Adams, J. B., Gregory, A. C., Borody, T., Chittick, L., Fasano, A., Khoruts, A., Geis, E., Maldonado, J., McDonough-Means, S., Pollard, E. L., Roux, S., Sadowsky, M. J., Lipson, K. S., Sullivan, M. B., Caporaso, J. G., & Krajmalnik-Brown, R. (2017). Microbiota Transfer Therapy alters gut ecosystem and improves gastrointestinal and autism symptoms: an open-label study. *Microbiome*, *5*(1), 10. https://doi.org/10.1186/s40168-016-0225-7
12. Desbonnet, L., Clarke, G., Shanahan, F. *et al.* Microbiota is essential for social development in the mouse. *Mol Psychiatry* 19**,** 146–148 (2014). https://doi.org/10.1038/mp.2013.65
13. Kaidanovich-Beilin, O., Lipina, T., Vukobradovic, I., Roder, J., & Woodgett, J. R. (2011). Assessment of social interaction behaviors. *Journal of visualized experiments : JoVE*, (48), 2473. https://doi.org/10.3791/2473
14. Gacias, M., Gaspari, S., Santos, P. M., Tamburini, S., Andrade, M., Zhang, F., Shen, N., Tolstikov, V., Kiebish, M. A., Dupree, J. L., Zachariou, V., Clemente, J. C., & Casaccia, P. (2016). Microbiota-driven transcriptional changes in prefrontal cortex override genetic differences in social behavior. *eLife*, *5*, e13442. https://doi.org/10.7554/eLife.13442
15. Kuntz, T., & Gilbert, J. (2016). Does the brain listen to the gut?. *eLife*, *5*, e17052. https://doi.org/10.7554/eLife.17052
16. Daneman, R., & Prat, A. (2015). The blood-brain barrier. *Cold Spring Harbor perspectives in biology*, *7*(1), a020412. https://doi.org/10.1101/cshperspect.a020412
17. Hoyles, L., Snelling, T., Umlai, U. K., Nicholson, J. K., Carding, S. R., Glen, R. C., & McArthur, S. (2018). Microbiome-host systems interactions: protective effects of propionate upon the blood-brain

barrier. *Microbiome*, *6*(1), 55. https://doi.org/10.1186/s40168-018-0439-y

18. Tillmann, S., Abildgaard, A., Winther, G., & Wegener, G. (2019). Altered fecal microbiota composition in the Flinders sensitive line rat model of depression. *Psychopharmacology*, *236*(5), 1445–1457. https://doi.org/10.1007/s00213-018-5094-2
19. Valles-Colomer, M., Falony, G., Darzi, Y. *et al.* The neuroactive potential of the human gut microbiota in quality of life and depression. *Nat Microbiol* **4,** 623–632 (2019). https://doi.org/10.1038/s41564-018-0337-x
20. Cheung, S. G., Goldenthal, A. R., Uhlemann, A. C., Mann, J. J., Miller, J. M., & Sublette, M. E. (2019). Systematic Review of Gut Microbiota and Major Depression. *Frontiers in psychiatry*, *10*, 34. https://doi.org/10.3389/fpsyt.2019.00034
21. Tremlett, H., & Waubant, E. (2018). The Gut Microbiota and Pediatric Multiple Sclerosis: Recent Findings. *Neurotherapeutics : the journal of the American Society for Experimental NeuroTherapeutics*, *15*(1), 102–108. https://doi.org/10.1007/s13311-017-0574-3
22. Haran, J. P., Bhattarai, S. K., Foley, S. E., Dutta, P., Ward, D. V., Bucci, V., & McCormick, B. A. (2019). Alzheimer's Disease Microbiome Is Associated with Dysregulation of the Anti-Inflammatory P-Glycoprotein Pathway. *mBio*, *10*(3), e00632-19. https://doi.org/10.1128/mBio.00632-19
23. Marino BLB, de Souza LR, Sousa KPA, Ferreira JV, Padilha EC, da Silva CHTP, Taft CA, Hage-Melim LIS. Parkinson's Disease: A Review from Pathophysiology to Treatment. Mini Rev Med Chem. 2020;20(9):754-767. doi: 10.2174/1389557519666191104110908. PMID: 31686637.
24. Gorecki, A. M., Preskey, L., Bakeberg, M. C., Kenna, J. E., Gildenhuys, C., MacDougall, G., Dunlop, S. A., Mastaglia, F. L., Akkari, P. A., Koengten, F., & Anderton, R. S. (2019). Altered Gut Microbiome in Parkinson's Disease and the Influence of Lipopolysaccharide in a Human α-Synuclein Over-Expressing Mouse Model. *Frontiers in neuroscience*, *13*, 839. https://doi.org/10.3389/fnins.2019.00839
25. Liu P, Peng G, Zhang N, Wang B and Luo B (2019) Crosstalk Between the Gut Microbiota and the Brain: An Update on Neuroimaging Findings. *Front. Neurol.* 10:883. doi: 10.3389/fneur.2019.00883

26. De Santis S, Moratal D, Canals S. Radiomicrobiomics: Advancing Along the Gut-brain Axis Through Big Data Analysis. Neuroscience. 2019 Apr 1;403:145-149. doi: 10.1016/j.neuroscience.2017.11.055. Epub 2017 Dec 10. PMID: 29237568.

27. Lin D, Ding J, Liu J-Y, He Y-F, Dai Z, Chen C-Z, et al. (2013) Decreased Serum Hepcidin Concentration Correlates with Brain Iron Deposition in Patients with HBV-Related Cirrhosis. PLoS ONE 8(6): e65551. https://doi.org/10.1371/journal.pone.0065551

28. Raimondi I, Izzo L, Tunesi M, Comar M, Albani D and Giordano C (2020) Organ-On-A-Chip *in vitro* Models of the Brain and the Blood-Brain Barrier and Their Value to Study the Microbiota-Gut-Brain Axis in Neurodegeneration. *Front. Bioeng. Biotechnol.* 7:435. doi: 10.3389/fbioe.2019.00435

29. Griep, L. M., Wolbers, F., de Wagenaar, B., ter Braak, P. M., Weksler, B. B., Romero, I. A., et al. (2013). BBB on chip: microfluidic platform to mechanically and biochemically modulate blood-brain barrier function. *Biomed. Microdevices* 15, 145–150. doi: 10.1007/s10544-012-9699-7

30. Zhu, S., Jiang, Y., Xu, K. *et al.* The progress of gut microbiome research related to brain disorders. *J Neuroinflammation* 17, 25 (2020). https://doi.org/10.1186/s12974-020-1705-z

Chapter 4

GUT MICROBIOTA AND OTHER DISEASES

"Treat our microbiome like a garden – we need to nourish the soil (intestines) for healthy plants (bacteria), while minimising weeds (disease-causing microbes.)"

- Tim Spector, a leading researcher in gut health

Shiv Sanjeevi Sripathi[1]

[1]Shiv Sanjeevi Sripathi secured second rank in Mumbai University and first rank in his college at his undergrad level and in his college during his Masters and has worked on a stem cell project at the Specialized Centre for Cell Based Therapy (SCCT), KEM Hospital at Mumbai, following which he was a Junior Research Fellow at Microbiology & Cell Biology Department, Indian Institute of Science, Bangalore from August 2009 to November 2011. The field of work involved cloning of cell wall genes and transcription factors in E. coli & M. smegmatis. This was followed by a stint as a Research and Development Executive at Transasia Biomedicals, Mumbai where he worked on patent writing and Immunoassays from 2011 till 2013. In 2013, he entered the teaching field where he has been tutoring biology and chemistry to the students of 11th and 12th grades along with English grammar and communication skills to people of all age groups. Shiv also loves helping people write scientific manuscripts and has co-edited several books on life sciences and aspects of biotechnology with Apple Academic Press.

This chapter offers a snapshot of how the gut microbiota is altered in several other conditions such as high blood pressure, cancer, and diabetes.

4.1 HYPERTENSION

According to researchers Mills and team (2016), the incidence of high blood pressure was 1.39 billion people globally with 349 million in high-income countries and 1.04 billion in low- and middle-income countries. Accounting for cardiovascular disease and early death, the uniformity in the trend of the global mean blood pressure (BP) can be accounted to the rife utilization of antihypertensive medications (Mills et al, 2020).

While multiple factors can be attributed to this condition ranging from genetics (specific loci in the renin-angiotensin–aldosterone system) to the lifestyle inclusive of compromised physical activity, alcohol, and dietary salt, research is pointing out how the gut microbiota is relevant in this global medical condition.

Researchers Yan and team (2017) reported the metagenomic profiling of patients with hypertension (≥140/90 mm Hg) vs. that of matched controls (n=60 each). An evident dysbiosis of the gut microbiota was revealed with the controls displaying an abundance of *Roseburia*, *Faecalibacterium,* and *Synergistetes* while the patients showed a higher count of *Klebsiella, Clostridium, Streptococcus, Parabacteroides, Eggerthella*, and *Salmonella*. Additionally, gene signature analyses revealed a variation in 53,953 genes between the groups that were presented as metagenome linkage groups (MLGs). The controls showed a predominance of *Faecalibacterium prausnitzii* and *Roseburia*-MLGs while the patients displayed enrichment of *Streptococcus* (*S. pasteurianus*, *S. infantarius* and *S. salivarius*), *Parabacteroides merdae*, *Klebsiella* (*K. pneumoniae* and *K. variicola*), *Pyramidobacter piscolens*, *Bacteroides* spp. (*B. cellulosilyticus* and *B. eggerthii*), and *Sutterella wadsworthensis*.

KEGG pathway analyses unveiled a depletion of SCFA-producing enzymes and an enrichment of trimethylamine (TMA: shown to augment atherosclerosis), synthesis of lipopolysaccharide (LPS) and membrane transport. Further, *Streptococcus* (*S. salivarius* and *S. infantarius*) were found to distinguish hypertensive patients from controls showing the utility of gut microbiota to predict hypertension.

The genera *Streptococcus, Clostridium*, and *Klebsiella* elevated in the patients with hypertension metabolize choline that is linked to an elevated risk of cardiovascular disorders. *Klebsiella* is associated with inflammatory gut diseases while *Streptococcus* is linked with cirrhosis and gut inflammation. The species of *Bacteroides* associated with the patients are linked to opportunistic infections.

Looking at the controls, *Roseburia* and *F. prausnitzii* are SFCA producers while *Megasphaera* synthesizes vitamins and SFCAs. This "hypertensive gut microbiome" is another angle that is being uncovered by research to look at targeting this global menace.

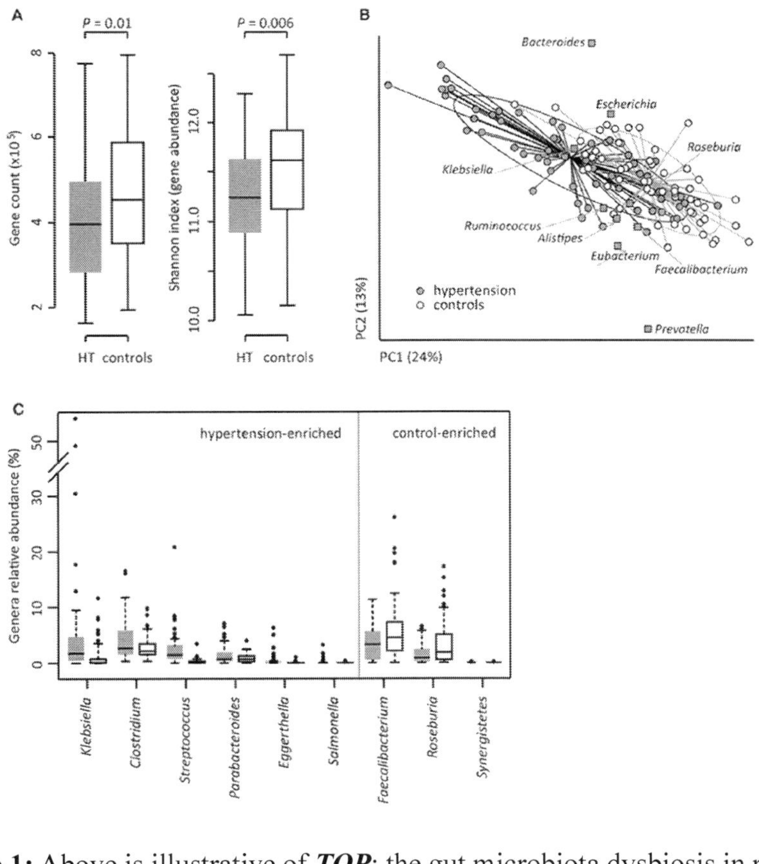

Figure 1: Above is illustrative of ***TOP***: the gut microbiota dysbiosis in patients with hypertension (HT) vs. controls: A) alpha diversity variation; the taxonomic variations are represented as B) the Bray-Curtis distances C) the genera shown as a Boxplot. Patients are gray boxes while controls are white boxes (2017 Yan, Gu, Li, Yang, Jia, Chen, Han, Huang, Zhao, Li, Fang, Zhou, Guan, Ding, Wang, Khan, Xin, Li and Ma. *This is an open-access article distributed under the terms of the Creative Commons Attribution License (CC BY). The use, distribution or reproduction in other forums is permitted, provided the original author(s) or*

licensor are credited and that the original publication in this journal is cited, in accordance with accepted academic practice. No use, distribution or reproduction is permitted which does not comply with these terms).

BOTTOM: Differences in the functions altered due to the dysbiosis. A) KEGG B) the controls and patients differing in SFCA and cutC (production of TMA) (2017 Yan, Gu, Li, Yang, Jia, Chen, Han, Huang, Zhao, Li, Fang, Zhou, Guan, Ding, Wang, Khan, Xin, Li and Ma. *This is an open-access article distributed under the terms of the* ***Creative Commons Attribution License (CC BY)****. The use, distribution or reproduction in other forums is permitted, provided the original author(s) or licensor are credited and that the original publication in this journal is cited, in accordance with accepted academic practice. No use, distribution or reproduction is permitted which does not comply with these terms*):

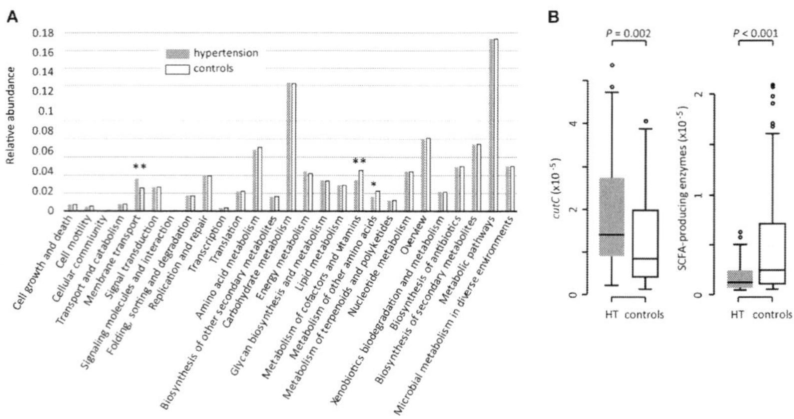

Scientists Li and team (2017) reported an analysis of the gut microbiota entailing metagenomics and metabolomics in patients sorted as per clinical guidelines of hypertension by the American Society of Hypertension and the International Society of Hypertension. The sample included 196 Chinese individuals: pre-hypertension (pHTN; n=59), hypertension (HTN; n=99) and controls (n=41). The pHTN and HTN samples showed a dip in the α-diversity at the genus level in the gut microbiota in lieu of earlier reports. This was followed by clustering the samples into enterotypes: enterotype 1 with a probable association in pHTN and HTN; enriched in *Prevotella* and enterotype 2 with a probable association in the controls; enriched in *Bacteroides*.

The controls, pHTN and HTN samples displayed a variation in 44 genera: *Prevotella* and *Klebsiella* were enriched in pHTN or HTN while HTN showed enrichment in *Porphyromonas* and *Actinomyces*. *Prevotella* is implicated in rheumatoid arthritis and periodontal diseases; *Klebsiella* with infections while *Porphyromonas* and *Actinomyces* with oral diseases. The levels of *Bifidobacterium* (that lowers LPS and maintains gut homeostasis), *Faecalibacterium* and *Roseburia* (both SFCA-butyrate producers) were lowered in the patients. Further, this dysbiosis was found to be associated evidently with metabolic changes in the host. These were inclusive of LPS production and export (associated with inflammation) along with phosphotransferase system (PTS), phosphatidylethanolamine, and lipid transport (linked to arthritis, cirrhosis and diabetes). Additionally, the levels of the anti-inflammatory molecules 3,4,5-trimethoxycinnamic acid and S-Carboxymethyl-L-cysteine were lowered in pHTN and HTN. Interestingly, a combination of the co-abundance groups (CAGs of the genes) and metabolites could distinguish controls from the PHTN and HTN groups but clear discrimination between these two groups was not achieved **suggestive of the likeness of pHTN and HTN**.

Transplantation of HTN patient's and control fecal samples into germ-free (GF) mice followed by 16S V4 region amplicon sequencing revealed that the HTN-group animals displayed lowered bacterial diversity. These HTN mice displayed an abundance of *Prevotella* and *Coprobacillus* and a dip in *Roseburia*, *Bifidobacterium*, *Ruminococcus*, etc. that was in lieu of the above analyses in patients. The systolic/diastolic and mean blood pressures were elevated in the HTN-sample transplanted mice, as determined by the tail-cuff method.

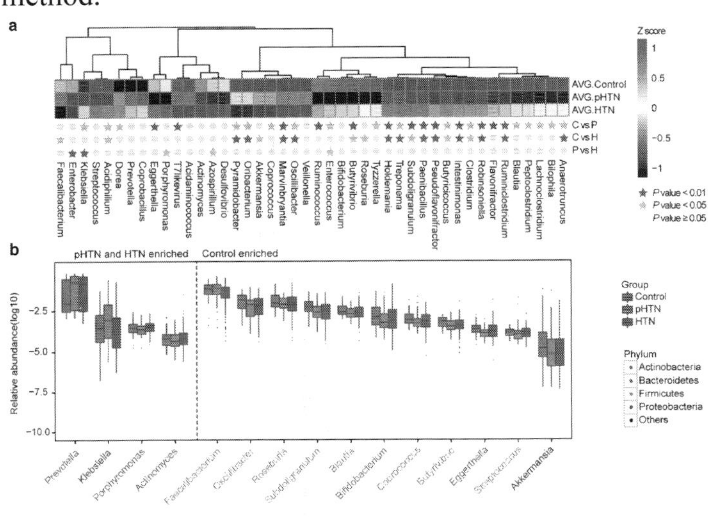

Figure 2: below displays the variation in genera ***TOP***: a) relative abundance: C, control; P, pHTN; H, HTN; dark green star: P-value <0.01; light green star:

P-value <0.05; gray circle: P-value ≥0.05 b) pHTN and HTN patients display enrichment in 4 genera while controls show 11 genera (Li et al, 2017: *Open-Access: This article is distributed under the terms of the Creative Commons Attribution 4.0 International License (http://creativecommons.org/licenses/by/4.0/), which permits unrestricted use, distribution, and reproduction in any medium, provided you give appropriate credit to the original author(s) and the source, provide a link to the Creative Commons license, and indicate if changes were made. The Creative Commons Public Domain Dedication waiver (http://creativecommons.org/publicdomain/zero/1.0/) applies to the data made available in this article, unless otherwise stated*).

BOTTOM: a) fecal microbiota transplantation from 2 HTN patients in 10 GF mice and 1 control individual in 5 GF mice b) Gut microbiome similarity in the donors c) HTN-transplanted mice with lowered α-diversity d) Difference in gut microbiota between HTN and control-transplanted mice e) Alterations shown as a heat map (less abundant in blue and more abundant in red) e) blood pressure readings in the mice (Li et al, 2017: *Open-Access: This article is distributed under the terms of the Creative Commons Attribution 4.0 International License (http://creativecommons.org/licenses/by/4.0/), which permits unrestricted use, distribution, and reproduction in any medium, provided you give appropriate credit to the original author(s) and the source, provide a link to the Creative Commons license, and indicate if changes were made. The Creative Commons Public Domain Dedication waiver (http://creativecommons.org/publicdomain/zero/1.0/) applies to the data made available in this article, unless otherwise stated*).

Overall, an aberration in the gut microbiota and their metabolites were unveiled in prehypertensive and hypertensive patients. The finding of hypertension involving dysbiosis of the gut bacteria opens up the avenue of adding the modulation of the gut microbes to the current approaches to treat hypertension (Li et al, 2017).

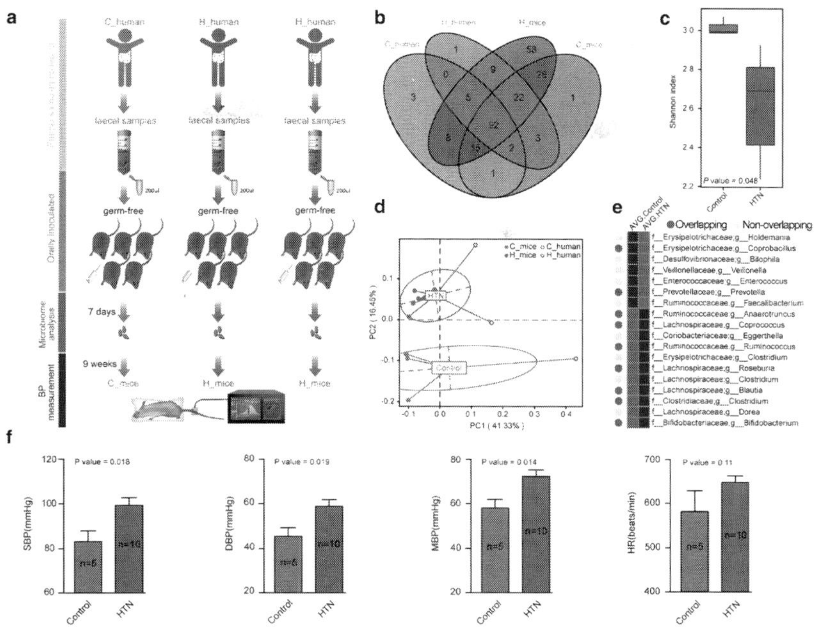

A recent report in 2020 by Silveira-Nunes and team analyzed the trend in the inflammatory cytokines along with gut microbiota in 32 controls and 48 hypertensive patients (defined as (higher than 140 mmHg for systolic and 90 mmHg for diastolic). The gut microbiota of the latter group displayed a lowering in Bacteroidetes and an elevation of the ratio of Firmicutes/ Bacteroidetes. Further, butyrate producers like *Faecalibacterium* and *Roseburia* were seen more in the controls than the patients. This becomes pertinent given that SFCAs are vitally involved in homeostasis and targeting of GPR41 (G protein-coupled receptor 41) and Olfr78 (Olfactory receptor 78) to control the blood pressure.

The cytometric bead array (CBA) for inflammatory cytokines revealed that the hypertensive patients displayed an elevated TNF-α/IFN-γ ratio that is suggestive of increased inflammation in these individuals. Analyses of the leukocyte profile revealed that half of the patients displayed high levels of TNF and IL-6. Research has cast light on the involvement of both these cytokines in hypertension. Also, the lowered *Roseburia* enrichment in patients becomes of relevance here given that these bacteria not only synthesize the anti-inflammatory SFCA but also conjugated linoleic acid that lowers the blood pressure.

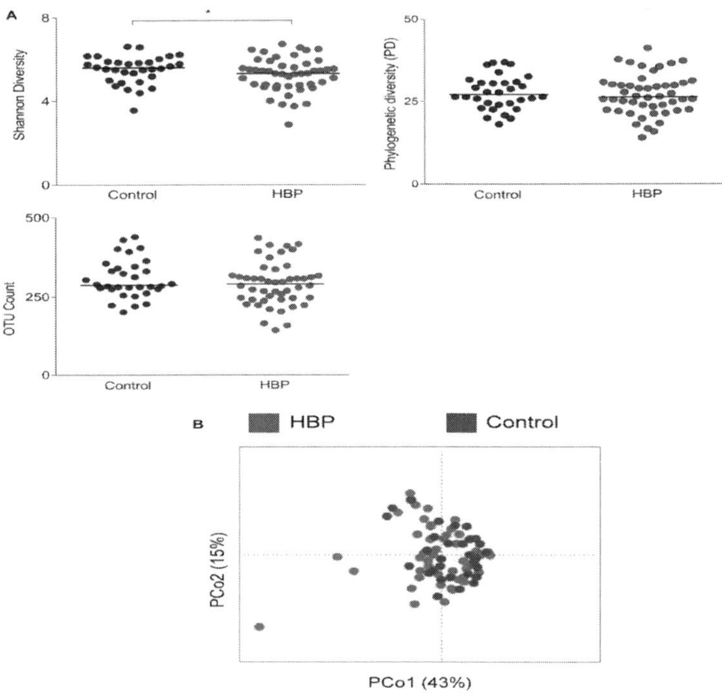

Figure 3: Above is illustrative of the dysbiosis of gut microbes in patients with hypertension (HBP) and controls ***TOP***: A) Alpha-diversity B) Principal Coordinates Analysis (2020 Silveira-Nunes, Durso, Alves de Oliveira, Cunha, Maioli, Vieira, Speziali, Corrêa-Oliveira, Martins-Filho, Teixeira-Carvalho, Franceschi, Rampelli, Turroni, Brigidi and Faria. *This is an open-access article distributed under the terms of the* ***Creative Commons Attribution License (CC BY)****. The use, distribution or reproduction in other forums is permitted, provided the original author(s) and the copyright owner(s) are credited and that the original publication in this journal is cited, in accordance with accepted academic practice. No use, distribution or reproduction is permitted which does not comply with these terms*).

BOTTOM A) Abundance at genus level B) Lachnospiraceae (Coprococcus and Roseburia) levels C) Variation in the genera as per Linear discriminant analysis (LDA); controls in blue and patients in red (2020 Silveira-Nunes, Durso, Alves de Oliveira, Cunha, Maioli, Vieira, Speziali, Corrêa-Oliveira, Martins-Filho, Teixeira-Carvalho, Franceschi, Rampelli, Turroni, Brigidi and Faria. *This is an open-access article distributed under the terms of the* ***Creative Commons Attribution License (CC BY)****. The use, distribution or reproduction in other forums is permitted, provided the original author(s) and the copyright owner(s) are credited and that the* *original publication in this journal is cited, in accordance with accepted academic practice. No use, distribution or reproduction is permitted which does not comply with these terms*):

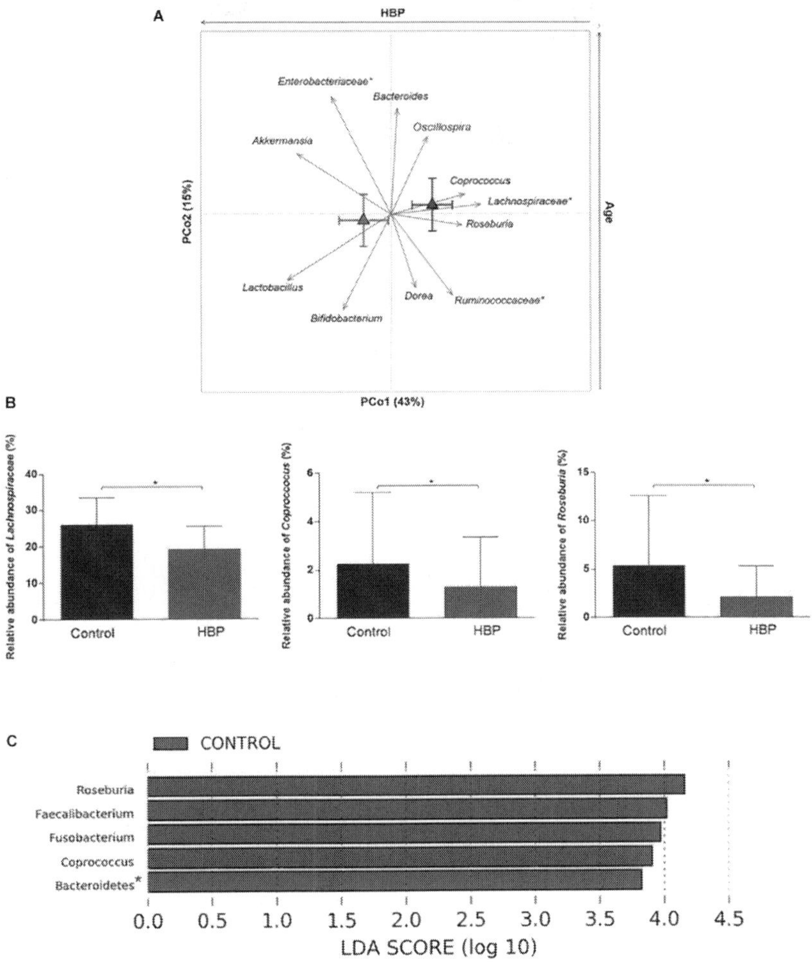

Overall, these few studies open up the avenue for further exploring the control of blood pressure using the modulation of the gut bacteria given the dysbiosis seen in this condition. This becomes extremely crucial given that 7.6 million deaths are caused globally due to elevated blood pressure that in turn causes 54% of stroke and 47% of coronary heart disease conditions (Arima et al, 2011).

4.2 THE ALTERED GUT MICROBES IN THE "SWEET" DISEASE

The occurrence of diabetes, as estimated by the International Diabetes Federation Diabetes Atlas stands at 463 million people or just below half a billion individuals globally. Urban and high-income areas displayed a higher predominance vs. that of rural and low-income areas. What is alarming is that one of two individuals is not aware that he/she has diabetes (Saeedi et al, 2019).

Not only is it a worrying cause of the global disease burden, type 2 diabetes (T2D) a primary risk factor for stroke and heart disease. Insulin-resistance, stress, a lack of physical activity, genetics play a combined role to manifest as this condition (Li et al, 2020).

2010-published research revealed "compositional changes in intestinal microbiota" in 18 individuals with T2D vs. that of 18 non-diabetic controls. V4 16S rRNA pyrosequencing showed that the bacterial diversity and richness were more in diabetics whose BMI exceeded 31 over lean individuals. The controls displayed enrichment of *Firmicutes* while the diabetics showed a little abundance of phylum *Bacteroidetes* and *Proteobacteria.* The *Bacteroidetes* to *Firmicutes* ratio displayed a negative correlation with BMI and a positive one with plasma glucose (as determined by the oral glucose tolerance test [OGTT]). *Roseburia* and *Prevotella* displayed negative and positive correlations, respectively with the plasma glucose. Principal component analysis (PCA) revealed a higher score for *Clostridia* and *Erysipelotrichi* vs. *Betaproteobacteria*, *Bacteroidetes* and *Bacilli* for the control vs. diabetic groups, respectively. *Prevotella*, *Lachnospiraceae IS*, *Roseburia* and *Subdoligranulum* also facilitated the distinguishing of diabetics vs. controls in some cases. This becomes pertinent given that the gram-negative bacteria *Bacteroidetes* and *Proteobacteria* synthesize LPS that is linked to gut inflammation.

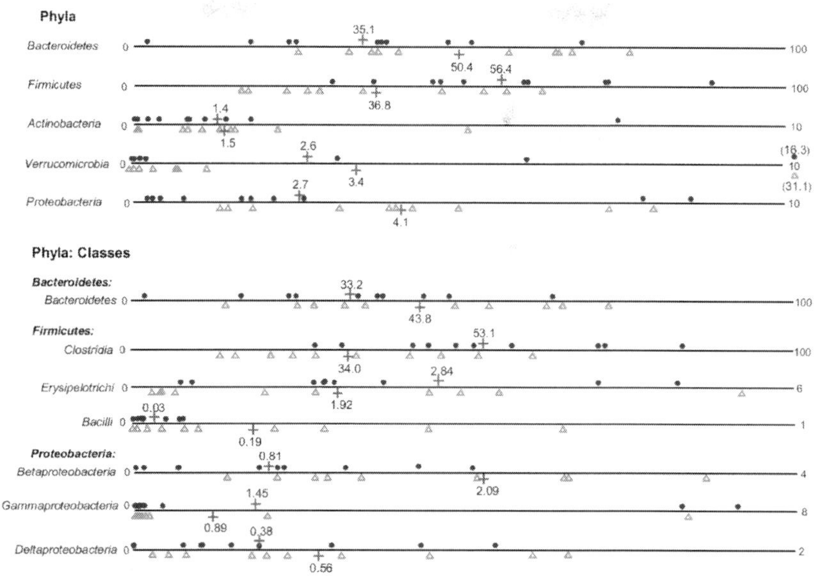

Figure 5: Above is illustrative of ***TOP***: bacterial abundances in T2D patients (green) vs. controls (blue); 2010 Larsen et al: *This is an open-access article distributed under the terms of the Creative Commons Attribution License, which permits unrestricted use, distribution, and reproduction in any medium, provided the original author and source are credited*).

BOTTOM: Abundances (relative) of A) *Bacteroidetes* to *Firmicutes* ratio B) *Clostridia* C) *Betaproteobacteria* D) BMI vs. *Bacteroidetes* to *Firmicutes* ratio (2010 Larsen et al: *This is an open-access article distributed under the terms of the Creative Commons Attribution License, which permits unrestricted use, distribution, and reproduction in any medium, provided the original author and source are credited*):

The manifestations of diabetes (as serum biomarkers) and the gut microbiota were evaluated in prediabetics (PreDMs), newly diagnosed diabetics, and diabetics vs, that of healthy controls by researchers Akshay and team (2020) in mSystems®, an open-access journal from the American Society for Microbiology. These biomarkers ranged from fasting glucose to the lipid profile to folic acid to amino acids [these are inclusive of branched-chain amino acids (BCAA) and aromatic amino acids (AAA): associated with insulin resistance] to the inflammatory cytokine, Interleukin 6 (IL-6), lipid peroxides and lipopolysaccharide (LPS). The sample was described as seventeen prediabetics (PreDMs), eleven newly diagnosed diabetics (NewDMs), thirty-nine diabetics receiving treatment and thirty-five controls.

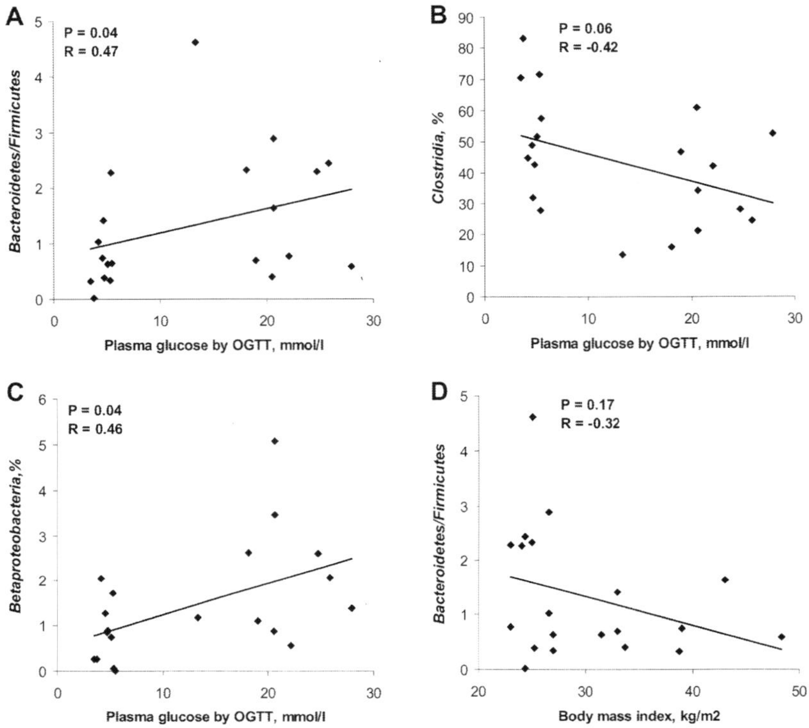

Diabetic individuals showed alterations in *Bacteroidetes*, *Firmicutes*, *Proteobacteria*, *Actinobacteria*, and *Verrucomicrobia*: with elevated *Firmicutes* and *Proteobacteria* and lowered *Bacteroidetes* in the recently diagnosed individuals. Compared to the controls *Verrucomicrobia* was lesser in NewDMs and *Actinobacteria* was more in the diabetics. The diabetics also showed lowered *Sutterella* while *Escherichia*, *Akkermansia, Megasphaera* and *Acidaminococcus* were elevated as opposed to the controls. The new DMs also showed elevated *Lactobacillus* while *Prevotella*, *Akkermansia*, *Blautia*, and *Ruminococcus* were lowered compared to controls. The bacterial diversity of PreDMs overlapped with that of the non-diabetics. Machine learning identified *Akkermansia* and *Sutterella* among the microbes to distinguish between the four groups assessed.

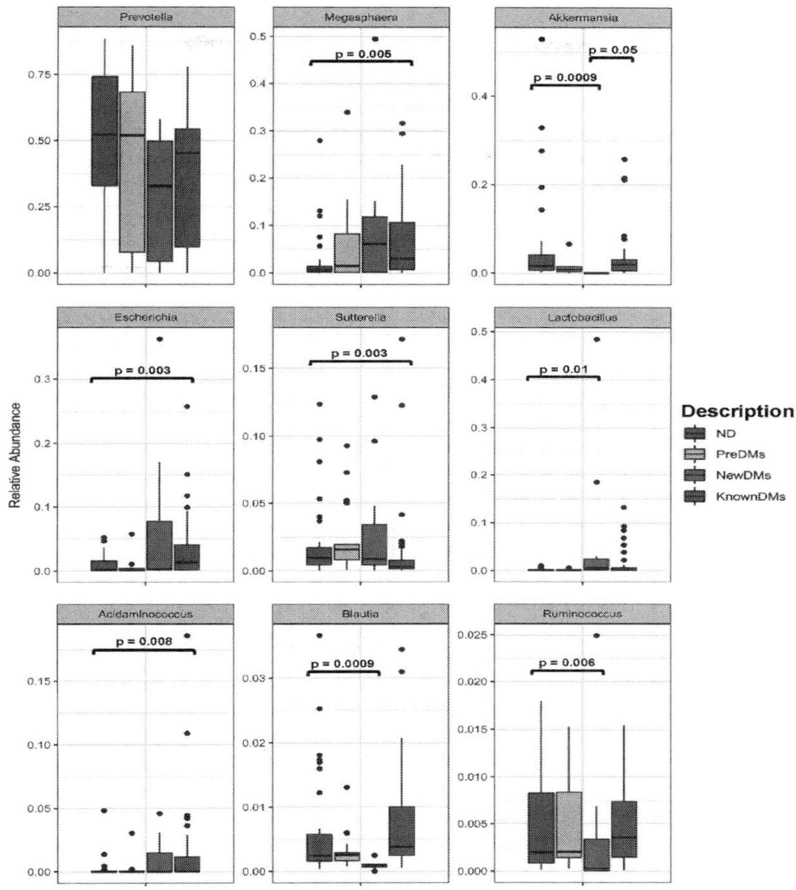

Figure 6: Above summarizes the findings of the analyses of ***TOP***: the genera abundant in controls, pre-diabetics, newly diagnosed diabetics and diabetics (© 2020 Gaike et al: *This is an open-access article distributed under the terms of the Creative Commons Attribution 4.0 International license. You are free to share — copy and redistribute the material in any medium or format; Adapt — remix, transform, and build upon the material for any purpose, even commercially*).

BOTTOM: Top ten discriminative (A) microbial genera. (B) serum biomarkers (© 2020 Gaike et al: *This is an open-access article distributed under the terms of the Creative Commons Attribution 4.0 International license. You are free to Share — copy and redistribute the material in any medium or format; Adapt — remix, transform, and build upon the material for any purpose, even commercially*):

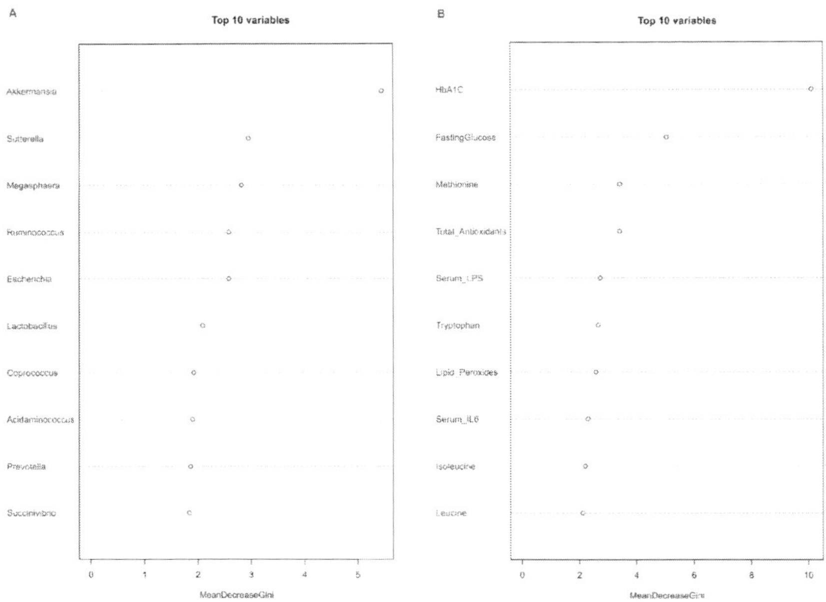

Interestingly, given the possible crucial involvement of rare bacteria (those that are less abundant) in the gut in metabolism, those types occurring below 0.01% in the total population were analyzed. The numbers of rare OTUs (operational taxonomic units) were more in the diabetics who were on medication.

Looking at the biochemical parameters and the microbes, the increased *Prevotella* displayed a positive correlation with histidine while *Akkermansia* and *Blautia* were positively linked to total antioxidants and histidine. The increased *Ruminococcus* was associated with histidine while *Megasphaera* was linked to HbA1c and fasting glucose. Fasting glucose, tyrosine, and lipid peroxides were positively correlated with the increased *Escherichia*. Abundant *Lactobacillus* was linked positively with HbA1c, fasting glucose and isoleucine. *Sutterella* was positively linked to HDL cholesterol and inversely linked to histidine and fasting glucose. *Acidaminococcus* was positively associated with fasting glucose, isoleucine, VLDL cholesterol, tyrosine, leucine and lipid peroxides.

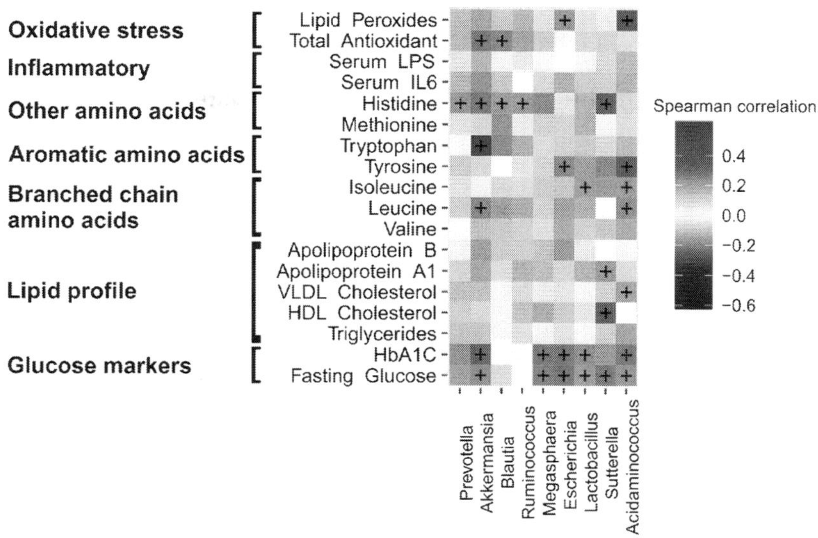

Figure 7: Above illustrates the associations between the altered gut bacteria and biochemical factors (© 2020 Gaike et al: *This is an open-access article distributed under the terms of the Creative Commons Attribution 4.0 International license. You are free to share — copy and redistribute the material in any medium or format; Adapt — remix, transform, and build upon the material for any purpose, even commercially*).

In another 2020-published report by Qian Li and team, the potential of employing gut microbiota as a tool to classify people with the risk of T2D development. 16S rRNA gene sequence and analyses of twenty people newly diagnosed with T2D and forty controls revealed elevated *Firmicutes* and lowered *Bacteroidetes* in the diabetic patients in lieu of literature. An increase was also seen for Fusobacterium in these subjects with lowered *Faecalibacterium*, *Bifidobacterium*, and *Akkermansia* (butyrate synthesizers). The gut bacteria in these T2D subjects also showed an interconnection, as outlined by coabundance network analysis. There was a positive association of *Rikenellaceae* **and** *Christensenellaceae* with *Butyricimonas* (known to target T2D) while negative connections were revealed for *Phascolarctobacterium and Dialister as well as Dorea* **with a negative correlation with** *Sutterella*.

Machine learning highlighted 50 discriminatory OTUs to facilitate the identification of controls vs. diabetics that showed an area under the receiver operating characteristic (ROC) curve (AUC) of 0.90, suggestive of the utility of this approach. The healthy gut was associated with *Bacteroides*, *Parabacteroides*, *Bifidobacterium*, *Oscillospira* and *Lachnospiraceae* one family) **while** *Clostridiales*, *Dorea*, *Faecalibacterium*, *Clostridiaceae* **and** *Lachnospiraceae* **(one family) were associated with the diabetic gut.**

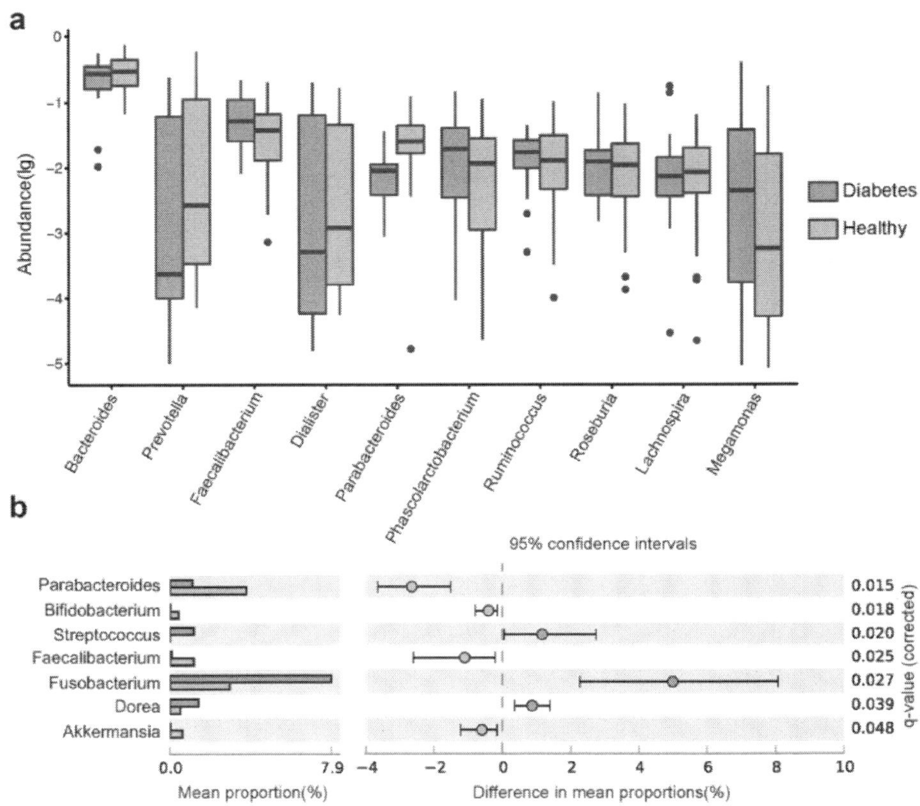

Figure 7: Above displays how the gut microbiota can be used to predict the risk of T2D. ***TOP***: a) bacterial genera abundance b) difference between the guts of T2D and control individuals: red is T2d; blue is control (Li et al, 2020: *Open-Access: This article is distributed under the terms of the Creative Commons Attribution 4.0 International License (http://creativecommons.org/licenses/by/4.0/), which permits unrestricted use, distribution, and reproduction in any medium, provided you give appropriate credit to the original author(s) and the source, provide a link to the Creative Commons license, and indicate if changes were made. The Creative Commons Public Domain Dedication waiver (http://creativecommons.org/publicdomain/zero/1.0/) applies to the data made available in this article, unless otherwise stated*):.

BOTTOM: The OTUs that emerge as predictors of control/T2D (Li et al, 2020: *Open-Access: This article is distributed under the terms of the Creative Commons Attribution 4.0 International License (http://creativecommons.org/licenses/by/4.0/), which permits unrestricted use, distribution, and reproduction in any medium, provided you give appropriate credit to the original author(s) and the source, provide a link to the Creative Commons license, and indicate if changes were made. The Creative Commons Public Domain Dedication waiver (http://creativecommons.org/publicdomain/zero/1.0/) applies to the data made available in this article, unless otherwise stated*):

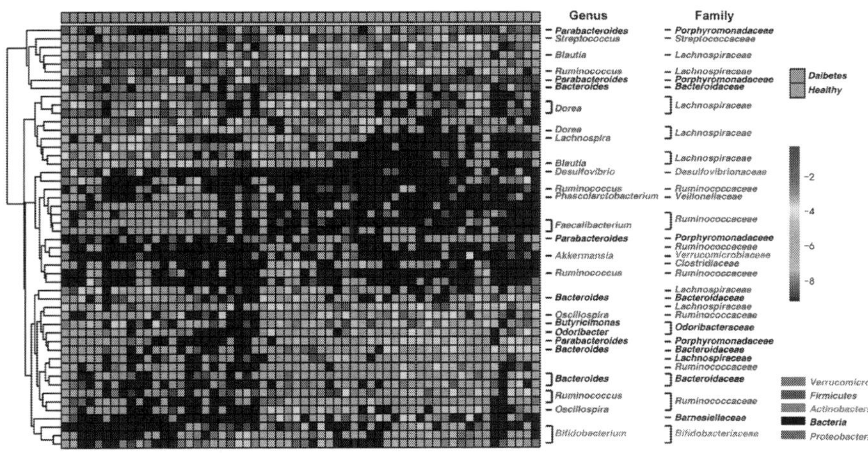

Let us look at the effect of **short-chain fatty acids (SCFAs)** produced by the gut microbiota. The major ones are acetate, propionate, and butyrate: the outcome of the gut bacteria processing dietary fiber and resistant starch. The signaling aspects of these molecules arise on account of interaction with free fatty acid receptors (acetate with FFRA2; propionate with FFRA2 or FFRA3; butyrate with FFRA3) (Puddu et al, 2014).

In a 2012-published work by Lin and team assessed the effects of these fatty acids on wild-type and mice with a knockout of *Ffar3*. The approach entailed the administration of a high-fat diet (HFD) along with butyrate (5% w/w), propionate (4.3%), and acetate (3.7%). The weight gain due to the HFD was inhibited by 40% by acetate while this gain of weight was entirely blocked by propionate and butyrate. The fasting glycemia was lowered in the animals administered propionate and the oral glucose tolerance was improved evidently by propionate and butyrate at the end of the term. All three SFCAs caused a dip in the levels of leptin and fasting insulin showing their effect on insulin sensitivity. The peptides associated with glucose metabolism: Peptide YY (PYY) and Glucagon-like peptide 1 (GLP-1) were stimulated in the order of butyrate followed by propionate and acetate.

Propionate and butyrate inhibited the weight gain and food intake on the same lines in the wild-type and mice with a knockout of *Ffar3*. Overall, the regulation of release of gut hormones and food intake by the SFCAs propionate and butyrate was displayed along with the ability of these gut microbiota-produced metabolites to exert protection against obesity/weight gain due to the diet (Lin et al, 2012).

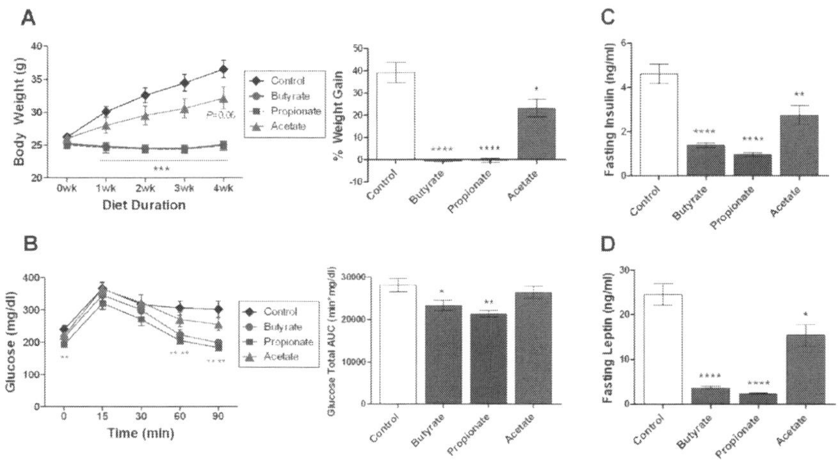

Figure 8: Above is illustrative of the effects of SFCAs on A) weight gain; B) Oral glucose tolerance; Plasma levels of C) insulin and D) leptin (2012 Lin et al, *This is an open-access article distributed under the terms of the Creative Commons Attribution License, which permits unrestricted use, distribution, and reproduction in any medium, provided the original author and source are credited*).

Research is uncovering the involvement of SCFAs in augmenting T2D-associated metabolism by binding to SCFA receptors in various tissues. For example, the inflammation associated with T2D is attributed to inflammatory molecules like tumor necrosis factor-alpha (TNF-α) and interleukin (IL)-6. The molecule Toll-like receptor 4 (TLR4) has also been shown to activate inflammatory cytokines to target the beta cells of the pancreas (Puddu et al, 2014).

Mattace Raso and team (2013) showed that administration of butyrate quelled the intensity of steatosis and liver inflammation in mice fed a high-fat diet (HFD). The expression of molecules associated with this inflammation like IL-6 and other molecules was attenuated by butyrate. The elevated TLR4 and its co-receptor CD14 expression in the liver of mice fed the HFD was conspicuously lowered by butyrate showing the potential of this molecule to address the inflammation associated with metabolic diseases.

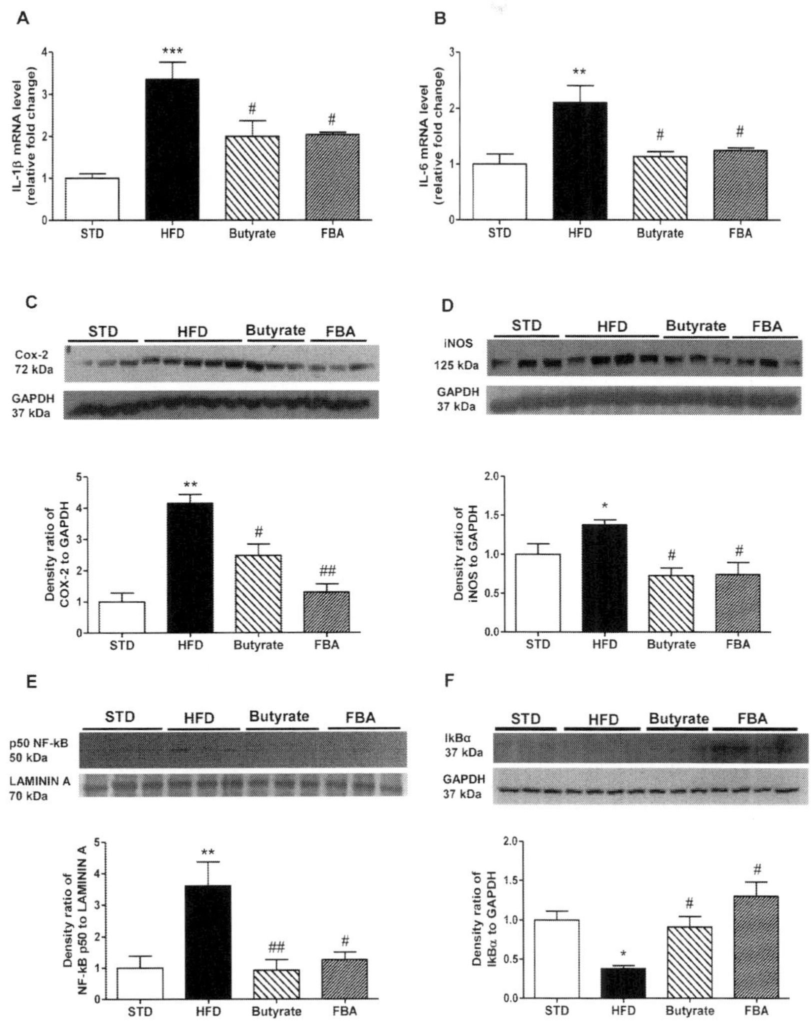

Figure 9: Above shows ***TOP***: the targeting of the elevated cytokines IL-1β (A) and IL-6 (B) and molecules COX-2 (C) and iNOS (D) by butyrate in the liver (2013, Mattace Raso et al: *This is an open-access article distributed under the terms of the Creative Commons Attribution License, which permits unrestricted use, distribution, and reproduction in any medium, provided the original author and source are credited*).

BOTTOM: lowered high–mobility group box 1 (HMGB1) (A) that activated TLR; TLR4 (B), TLR2 (C) and TLR9 (D) (2013, Mattace Raso et al: *This is an open-access article distributed under the terms of the Creative Commons Attribution License, which permits unrestricted use, distribution, and reproduction in any medium, provided the original author and source are credited*):

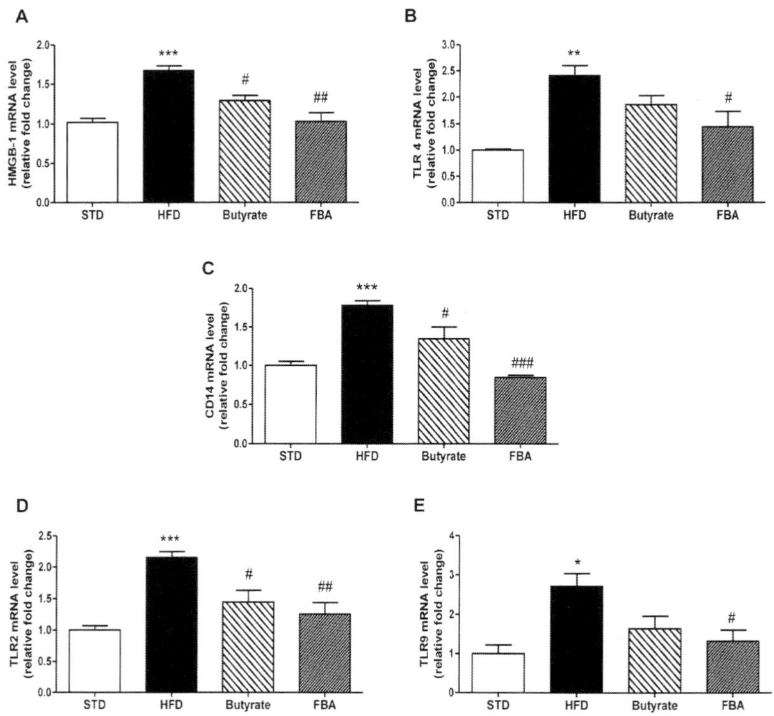

Further, propionate was found to target the release of the inflammatory TNF-α from neutrophils in response to LPS and also in adipocytes. Thus, SFCAs are associated with control of blood glucose levels in T2D via: a) targeting of the inflammation that causes insulin resistance, b) insulin release via GLP-1 release, c) protecting the beta cells of the pancreas to restore optimal glucose metabolism. Further research can add details and new insights into how the gut microbiota-mediated SFCAs emerge as a healer of aberrant glucose profiles in T2D (Puddu et al, 2014).

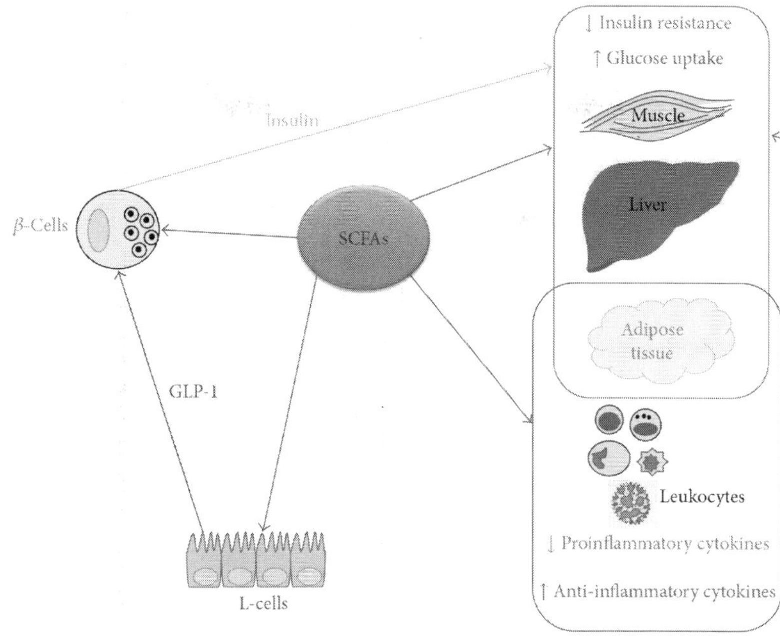

Figure 10: Above is illustrative of how SFCAs can quell the aberrant manifestations accounting for T2D (Alessandra Puddu et al, 2014: *This is an open-access article distributed under the Creative Commons Attribution License, which permits unrestricted use, distribution, and reproduction in any medium, provided the original work is properly cited*).

Scientists Zhang et al (2012) assessed the changes in the gut microbiota in a rat model in response to administration of the alkaloid berberine from *Coptis chinensis* demonstrated to be effective against bacterial infections and T2D as well as addressing the pro-inflammatory cytokine profile. The molecule resulted in lowering the expression of lipopolysaccharide (LPS)-binding protein (LBP) attributed to a possible dip in the antigen load. Further, the abundance of SFCA producers and SFCA levels were elevated with certain microbes like *Allobaculum* and *Blautia* reaching ten times that of the control rats fed only an HFD. **This opens up the avenue of employing gut bacteria modulation to address various manifestations of T2D!**

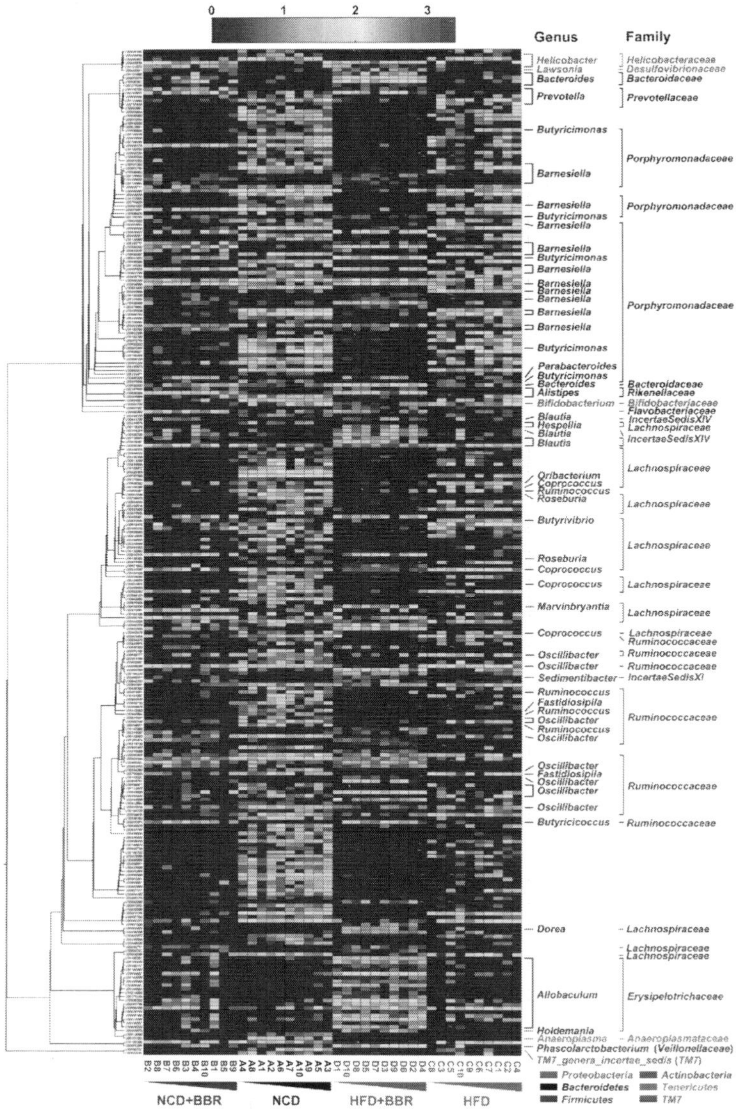

Figure 11: Above displays the altered gut microbiota profile such as that of the SFCA producers *Blautia* and *Allobaculum* by the alkaloid berberine for addressing T2D as a heat map. NCD: Normal chow diet; HFD" high-fat diet and BBR: berberine (Zhang et al, 2012: *This is an open-access article distributed under the terms of the Creative Commons Attribution License, which permits unrestricted use, distribution, and reproduction in any medium, provided the original author and source are credited*).

A 2019-published review discussed the potential of using herbs to address the dysbiosis of the gut microbiota in T2D. For example, *Plantago asiatica*

L seeds augmented the levels of the beneficial gut bacteria *Lactobacillus fermentum, Bacteroides*, and *Prevotella*. Further examples to just add flavor to how the gut microbiota can be modulated for addressing disease is presented below.

Table 1: shows gut microbiota regulation by herbs or their extracts to manage the "sweet disease" (2019, Boxun Zhang et al: *This is an open-access article distributed under the Creative Commons Attribution License, which permits unrestricted use, distribution, and reproduction in any medium, provided the original work is properly cited*):

Key: NEFA: nonesterified fatty acid; F/B: Firmicutes/Bacteroidetes; GLP-2: glucagon-like peptide-2; GLP-1: glucagon-like peptide-1; SCFAs: short chain fatty acids; N/A: not applicable.

Herbs/Herbal extracts	Models	Changes in the gut microbiota and their metabolites	Core mechanisms
Folium Mori	SD rats	**Increased**: the phyla Bacteroidetes and Proteobacteria and class Clostridia	Improve NEFA metabolism
Dendrobium candidum	KM mice	**Increased**: the abundance and diversity of gut microbiota	Attenuate oxidative stress
Rhizoma Dioscoreae	Human	**Increased**: *Bifidobacterium*	N/A
Berberine	Wistar/SD rats	**Increased**: *Bifidobacterium, Lactobacillus* **Decreased**: *Escherichia coli, Enterococcus*	Protect intestinal barrier; suppress inflammatory response; promote GLP-2 secretion
Rhein	db/db mice	**Increased**: Bacteroidetes, *Akkermansia* **Decreased**: Firmicutes, F/B ratio	Promote GLP-1 secretion
Seabuckthorn Protein	ICR mice	**Increased**: *Bifidobacterium, Lactobacillus, Bacteroides* **Decreased**: *Clostridium coccoides*, PH value	Improve fatty acids metabolism
Zanthoxylum alkylamides	SD rats	**Increased**: *Lactobacillus, Bifidobacterium, Clostridium*, SCFAs **Decreased**: *Enterococcus, Enterobacterium, Bacteroides* , PH value, free ammonia	Inhibit hepatic gluconeogenesis; promote insulin secretion

Polyphenols from *Fructus Mori L.*	db/db mice	**Increased**: Bacteroidetes, Anaeroplasmatales, butyrate, propionate **Decreased**: Firmicutes, F/B ratio, *Bacillus*, *Lactobacillus*	Attenuate oxidative stress
Polysaccharide from the seed of *Plantago asiatica* L.	Wistar rats	**Increased**: *Bacteroides*, *Lactobacillus*, *Prevotella*, SCFAs **Decreased**: *Alistipes obesi*	Improve NEFA metabolism; attenuate oxidative stress
Ganoderma atrum Polysaccharide	Wistar rats	**Increased**: SCFAs	Attenuate oxidative stress
Maydis stigma Polysaccharide	KM mice	**Increased**: *Lactobacillus*, *Bacteroides*	N/A
Radix Pseudostellariae Polysaccharide	C57 mice	**Increased**: *Lactobacillus*, *Bifidobacterium*	Attenuate oxidative stress; suppress inflammatory response
Morus nigra Polysaccharide	db/db mice	**Increased**: Bacteroidales, *Bacteroides Lactobacillus*, *Allobaculum*, *Akkermansia* **Decreased**: F/B ratio, opportunistic pathogens such as *Staphylococcus* and *Enterococcus*	Regulate lipid metabolism; attenuate oxidative stress
Polysaccharide from the fermented *Momordica charantia L*	Wistar rats	**Increased**: *Lactococcus laudensis*, *Prevotella loescheii*, SCFAs **Decreased**: pH value	Attenuate oxidative stress

Table 2: shows gut microbiota regulation by compounds from herbs to manage the "sweet disease" (2019, Boxun Zhang et al: *This is an open-access article distributed under the Creative Commons Attribution License, which permits unrestricted use, distribution, and reproduction in any medium, provided the original work is properly cited*):

Key: SCFAs: short-chain fatty acids; GLP-1: glucagon-like peptide-1; GLP-2: glucagon-like peptide-2; LPS: lipopolysaccharides; N/A: not applicable.

Preparations	Composition of preparations	Models	Changes of the gut microbiota and metabolites	Core mechanisms
Xiexin Tang	Rhizome Rhei, Radix Scutellaria, Rhizome Coptidis	SD rats	**Increased**: some SCFAs producing and anti-inflammatory bacteria such as *Adlercreutzia* and *Alloprevotella* ; SCFAs	Suppress inflammatory response

Gegen Qinlian Decoction	Radix Puerariae, Radix Scutellariae, Rhizoma Coptidis, Honey-fried Licorice Root	human	**Increased**: *Faecalibacterium*, *Bifidobacterium*, *Gemmiger* **Decreased**: *Alistipes*, *Odoribacter*	N/A
		KK-Ay mice	**Increased**: *Lactobacillus johnsonii*, *Bacteroides vulgates*	Suppress inflammatory response
AMC	Rhizoma Anemarrhenae, Momordica charantia, Rhizoma Coptidis, Salvia miltiorrhiza, Red yeast rice, Aloe vera, Schisandra chinensis, Rhizoma zingiberis	human	**Increased**: *Faecalibacterium*, *Roseburia*, *Gemmiger*, *Coprococcus*, *Megamonas*, *Blautia*	N/A
Huanglian Jiedu Decoction	Coptidis Rhizoma, Scutellariae Radix, Phellodendri Cortex, Gardeniae Fructus	ZDF rats	**Increased**: *Bacteroides*, *Clostridium*, *Roseburia*, *Akkermansia*, *Oscillospira*, *Aggregatibacter*, *Eubacterium* **Decreased**: *Prevotella*	Protect intestinal barrier; suppress inflammatory response; promote GLP-1and GLP-2 secretion
		SD rats	**Increased**: *Parabacteroides*, *Blautia*, *Akkermansia* **Decreased**: *Aerococcus*, *Staphylococcus*, *Corynebacterium*	Regulate lipid metabolism; suppress inflammatory response; attenuate oxidative stress
Qijian mixture	Astragalus membranaceus, Ramulus euonymi, Coptis chinensis, Pueraria lobata	KKay mice	**Increased**: Bacteroidetes	Improve carbohydrate and amino acid metabolism disorder
Banxia Xiexin Decoction	Rhizoma Pinelliae Praeparata, Radix Scutellaria, Rhizome Coptidis, Panax ginseng, Rhizoma zingiberis, Radix liquiritiae, Fructus Ziziphi Jujubae	SD rats	**Increased**: *Bacteroides*, *Bifidobacterium*, *Lactobacilli*, *Enterococcus*	Suppress inflammatory response
Herbal extract + Chemical drugs	Houttuynia cordata Thunb., Metformin	OLETF rats	**Increased**: *Roseburia*, *Akkermansia*, SCFAs **Decreased**: Gram-negative bacteria, *prevotella*, *Escherichia coli*, LPS	Suppress inflammatory response
Herbal extract +Probiotics	Burdock Fructooligosaccharide GF13 *Lactobacillus plantarum* Sc 52	C57 mice	**Increased**: *Lactobacillus*, *Bifidobacterium*, SCFAs **Decreased**: *Enterobacterium*, *Enterococcus*	Protect intestinal barrier; suppress inflammatory response;
Herbal extract +Prebiotics	Berberine, stachyose	KKay mice	**Increased**: *Lactobacillus*, *Bifidobacterium*	N/A

4.3 THE GUT MICROBES IN POLYCYSTIC OVARY SYNDROME (PCOS)

Polycystic ovary syndrome (PCOS) is observed in around 6-20% of women of reproductive ages and involves disruption of the hypothalamic-pituitary–

ovarian (HPO) axis vital for reproductive competency. The manifestations range from erratic menses to infertility and hirsutism. The disease is described as complex with the number of factors involved increasingly unveiled by research (Witchel et al, 2019).

Lindheim and team (2017) evaluated the gut microbiome in premenopausal women with PCOS and hormonally healthy controls (n=25 each). Analyses of the stool microbiome revealed a dip in the OTU count and a lowering of Faith's phylogenetic diversity by 15% in the PCOS patients vs. the controls. The abundance of the phylum Tenericutes, the order ML615J-28 (phylum Tenericutes), and the family S24-7 (phylum Bacteroidetes) was lower in PCOS patients. A dip in anti-Muellerian hormone (AMH), serum androgens and oligo-amenorrhoea (infrequency of menses) was associated with probands of these aforementioned bacteria. This pilot study cast light on the variation in phylogenetic diversity of gut bacteria that was connected to clinical factors of PCOS opening the avenue for deeper research.

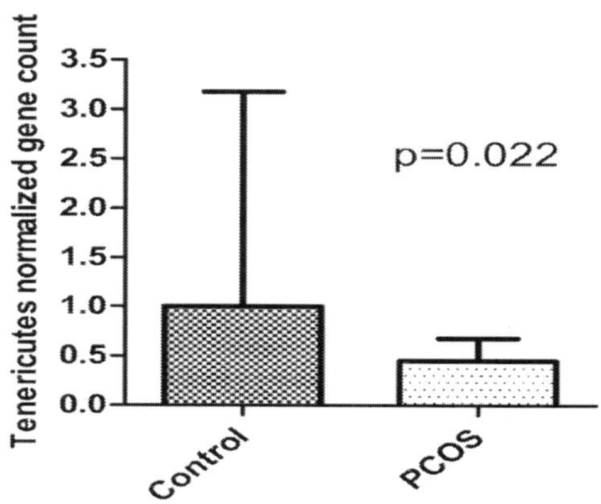

Figure 12: Above is illustrative of ***TOP***: Tenericutes and their lower abundance in PCOS patients (2017, Lindheim et al: *This is an open-access article distributed under the terms of the Creative Commons Attribution License, which permits unrestricted use, distribution, and reproduction in any medium, provided the original author and source are credited*).

BOTTOM: The altered gut microbiota abundance and association with diversity (A and B); serum androgens, AMH and oligo-amenorrhoea (C to H) and a lack of a link with hirsutism (I) (2017, Lindheim et al: *This is an open access article distributed under the terms of the Creative Commons Attribution License, which permits unrestricted use, distribution, and reproduction in any medium, provided the original author and source are credited*):

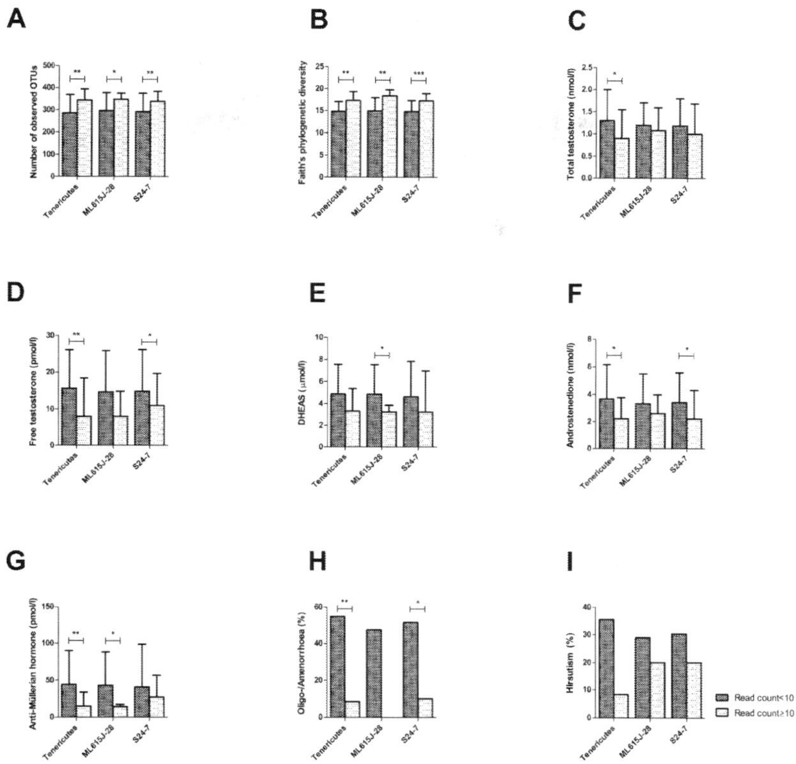

Looking at possible treatment in the future, researchers Guo and team (2016) assessed the gut microbiome of PCOS-induced rats vs. controls. The gut microbes were significantly altered in the rats with PCOS as revealed by Principal component analysis (PCA). The controls and PCOS groups displayed an alteration in seven bands in Denaturing gradient gel electrophoresis (DGGE) that were found to be *Pseudomonas monteilii SB3101*, *Roseburia intestinalis XB6B4*, *Prevotella melaninogenica ATCC25845*, *Prevotella denticola F0289*, *Lactobacillus johnsonii NCC533* **and** *Ruminococcus torques L2-14*. The PCOS rats displayed a lowering of *Lactobacillus*, *Ruminococcus* **and** *Clostridium* with an elevated profile of *Prevotella*. PCOS rats were further subjected to fecal microbial transplantation (FMT) or *Lactobacillus* **isolated from the controls.**

While the control rats displayed a normal ovarian appearance, the follicles were degenerating in the PCOS group as per morphological analyses. The regular estrous cycles of 4–5 days in the controls were contrasted with that of the PCOS animals that were always in the diestrus stage. The FMT resulted in an improvement of this aberrant estrous cycle. The PCOS rats that received FMT or *Lactobacillus* **transplantation displayed the formation**

of corpora lutea and granulose layers. The lowered concentrations of estradiol and estrone observed in the PCOS rats were increased by the gut microbiota transplantation with the efficiency of the FMT at par with that of the controls. Additionally, elevated testosterone and androstenedione **levels in the PCOS mice were also brought down conspicuously due to the FMS transplantation.**

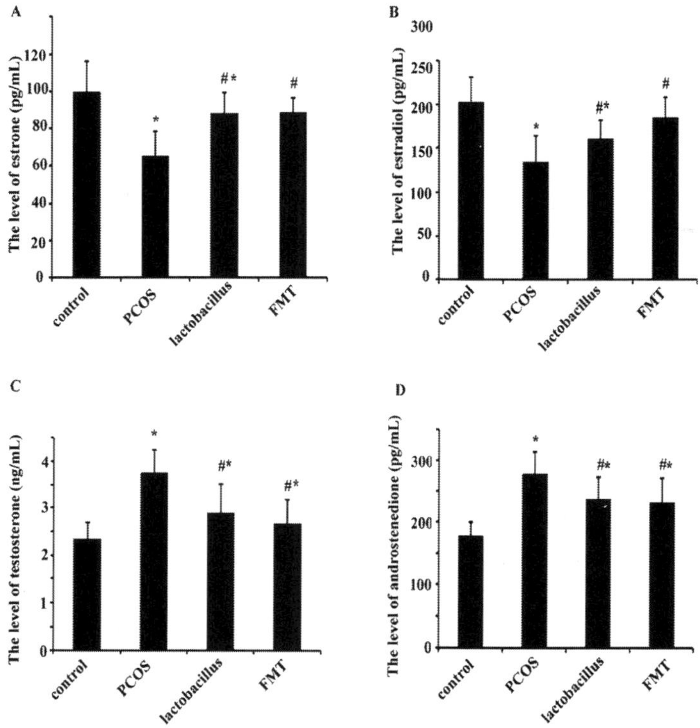

Figure 13: Above is illustrative of the ***TOP***: Improved hormone profile in PCOS rats in the groups: estrone (A), estradiol (B), androstenedione (C) and testosterone (D) (2016, Guo et al. *This is an open-access article distributed under the terms of the Creative Commons Attribution License, which permits unrestricted use, distribution, and reproduction in any medium, provided the original author and source are credited*).

BOTTOM: changes in the ovary of (A) Control group showing normal appearance. (B) PCOS group displays cystic degenerating follicles with thin granulosa layers. (C) PCOS Lactobacillus transplantation group increased granulosa layers. (D) PCOS FMT group showed increased granulosa layers and formation of corpora lutea. The larger boxed area in A, B, C, D was shown at higher magnification (400×) in E, F, G, H, respectively.

GLC: granular cell layer, TCL: theca cell layer, O: oocyte, L: luteum, AF: atretic follicle. Magnification (100×), scale bar, 50 μm (2016, Guo et al. *This is an open-access article distributed under the terms of the Creative Commons Attribution License, which permits unrestricted use, distribution, and reproduction in any medium, provided the original author and source are credited*):

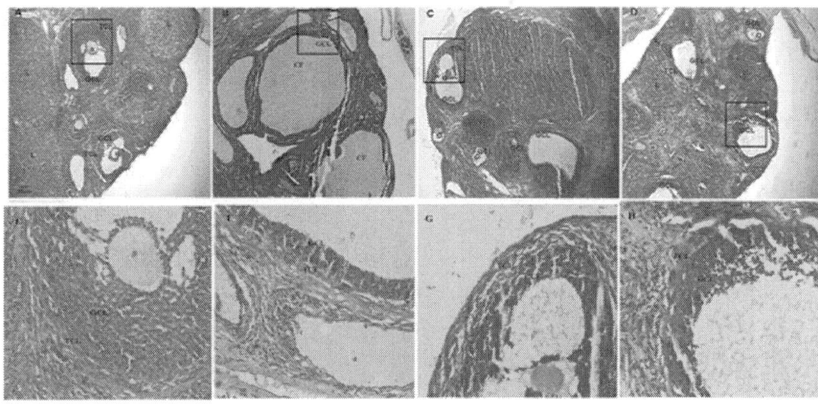

Overall, the improved steroid hormone profile and ovarian morphology due to gut microbiota open up the avenue to explore the addressing of this condition using our gut inhabitants.

4.4 THE BIG "C"

Research has shown that the presence of few microbiota members can trigger oncogenesis. An example is *Helicobacter pylori* that produces cytotoxin associated gene A (CagA), which leads to p53 destruction and hence, taking the risk of gastric cancer higher. DNA damage due to reactive oxygen species (ROS) can manifest as a result of activated spermine oxidase in the host cell by *Helicobacter pylori* or *Bacteroides fragilis.* The products of *Fusobacterium nucleatum* and enterotoxigenic *Bacteroides fragilis* (ETBF) can interact with E-cadherin to cause the expression of *c-myc* (a proto-oncogene). Tumorigenesis can be augmented by the secretion of growth factors by senescent cells, a phenomenon called senescence-associated secretory phenotype (SASP), which is triggered by colibactin-producing (*pks+) Escherichia coli*. Mice models have revealed that genomic instability is caused by *Enterococcus faecalis* infection that has also been reported for *in vitro* studies with *pks+ E. coli*. While these studies have shown the effect of modulation of host systems, an important point is that the composition and functions of the gut microbiota are shaped by the diet and lifestyle (and hence the immune system) of the host (Fulbright et al, 2017; Vivarelli et al, 2019).

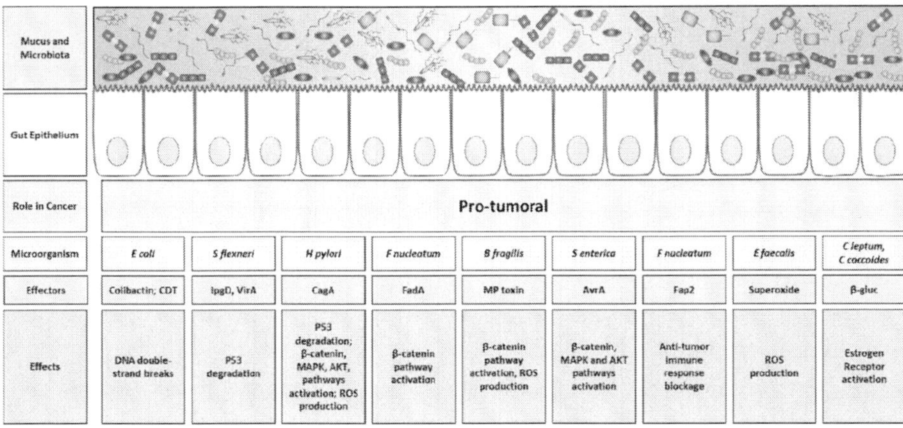

Figure 14: Above is demonstrative of how gut dysbiosis can support oncogenesis: *ROS, Reactive Oxygen Species; CTD, cytolethal distending toxin; IpgD, inositol phosphate phosphatase D; VirA, virulence gene A; CagA, cytotoxin associated gene A; FadA, Fusobacterium effector adhesin A; MP Toxin, metalloproteinase toxin; AvrA, avirulence protein A; β-gluc, β-glucuronidase* (Vivarelli et al, 2019: *Licensee MDPI, Basel, Switzerland. This article is an open-access article distributed under the terms and conditions of the Creative Commons Attribution (CC BY) license (http://creativecommons.org/licenses/by/4.0/)*.

Looking at the ability of gut bacteria to target cancer, tumor cell apoptosis has been shown to be triggered by ferricrome , a product from *Lactobacillus casei*. Further, the stimulation of a Th1-response or NK cell activity by Lactobacilli has also been outlined in studies. Interestingly, the recurrence of bladder cancer was lowered by *Lactobacillus casei* injection. The anti-inflammatory attributes of *Lactobacillus rhamnosus GG* (LGG) have propelled several studies to help maintain the gut integrity and the intestinal epithelial barrier (Vivarelli et al, 2019).

Table 3: below lists few clinical studies utilizing gut microbiota in cancer therapy: * Registered at ClinicalTrials.gov; C, Closed; O, Ongoing; GI, Gastrointestinal; DS, Dietary Supplement; CRC, Colorectal Cancer; CC, Colon Cancer; RC, Rectal Cancer; BC, Breast Cancer; NSCLC, Non-Small Cell Lung Cancer; PC, *Periampullary Carcinoma; LC, Lung Cancer; HCC, Hepatocellular Carcinoma; AML, Acute Myeloid Leukemia*

ClinicalTrials. Gov Identifier	Status	Title	Intervention	Disease
NCT00936572	C	Probiotics in CRC patients	DS: probiotic *Lal*	CRC
NCT01839721	C	Impact of probiotics on diarrhea in patients treated with pelvic radiation	DS: probiotic *Bifilact*	Various Cancers
NCT01410955	C	Prevention of irinotecan-induced diarrhea by probiotics	DS: probiotic *Colon Dophilus*	CRC
NCT01479907	C	Synbiotics and GI function-related quality of life after colectomy for cancer	DS: prebiotics and probiotics *Synbiotic Forte*	CRC
NCT01609660	C	Impact of probiotics on the intestinal microbiota	DS: *S boulardii*	CRC
NCT03072641	C	Using probiotics to reactivate tumor-suppressor genes in CRC	DS: probiotic *ProBion Clinica*	CC
NCT01468779	C	Effect of probiotics in patients undergoing surgery for periampullary neoplasms	DS: probiotics	PC
NCT01895530	C	Impact of probiotics in modulation of intestinal microbiota	DS: *S boulardii*	CRC
NCT03420443	C	Action of synbiotics on irradiated GI mucosa in RC treatment (FIPIREX)	DS: probiotics	RC
NCT02771470	C	Intestinal microbiota in lung cancer after chemotherapy	DS: probiotics	LC

NCT02021253	C	Influence of probiotics administration before liver resection in liver disease (LIPROCES)	DS: probiotics	HCC
NCT02751736	O	The effect of probiotics on bowel function restoration after ileostomy closure in patients with RC	DS: probiotic *CJLP243*	RC
NCT03290651	O	Probiotics and breast health	DS: probiotic *RepHresh Pro-B*	BC
NCT03518268	O	*Vivomixx* for prevention of bone loss in women with BC treated with an aromatase inhibitor	DS: probiotic *Vivomixx*	BC
NCT03177681	O	The effect of yogurt in cancer patient with moderate GI symptoms	DS: probiotics in yogurt	Various Cancers
NCT03642548	O	Probiotics combined with chemotherapy for patients with advanced NSCLC	Drug with DS of probiotic *Bifico*	NSCLC
NCT03358511	O	Engineering gut microbiome to target BC	DS: Probiotic *Primal Defense Ultra*	BC
NCT02944617	O	Probiotic yogurt supplement in reducing diarrhea in patients with metastatic kidney cancer being treated with VEGF-TK inhibitor	DS: probiotics in yogurt	Renal Cell Cancer
NCT02351089	O	Probiotics in radiation-treated gynecologic cancer (ProRad)	DS: probiotics	Gynecologic Cancer
NCT03574051	O	Microbiota are associated with Iodine-131 therapy and hypothyroidism	Iodine-131 Therapy with DS of probiotics	Thyroid Cancer
NCT03552458	O	Effects of Probiotics in preventing oral mucositis in patients undergoing head and neck radiotherapy	DS: probiotic *L Reuteri*	Head-and-neck Cancer
NCT02819960	O	prevention of irinotecan-induced diarrhea by probiotics	DS: probiotic *Probio-Fixinum* (including LGG)	CRC

NCT01790035	O	Probiotic LGG for prevention of side-effects in patients undergoing chemoradiation for GI cancer	DS: probiotic LGG	GI Cancer
NCT00197873	O	*Lactobacillus Rhamnosus* in prevention of chemotherapy-related diarrhea	DS: probiotic LGG	CRC
NCT02770326	O	Safety of stool transplant for patients with difficult to treat *C. difficile* infection	FMT	Various Cancers
NCT02928523	C	Prevention of dysbiosis complications with autologous FMT in acute myeloid leukemia patients undergoing intensive treatment (ODYSSEE)	Autologous FMT	AML
NCT03353402	O	FMT in metastatic melanoma patients who failed immunotherapy	FMT	Melanoma
NCT03341143	O	FMT in melanoma patients	FMT with Pembrolizumab	Melanoma

In mice models, combining *Lactobacilli* with cisplatin caused the induction of apoptosis in the tumor environment to improve the outcome of lung cancer. The involvement of the gut microbiota in immunotherapy was outlined in a study where the melanoma was controlled by orally administering *Bifidobacterium* at par with checkpoint blockade therapy employing programmed cell death protein 1 ligand 1 (PD-L1). Similarly, the efficacy of cyclophosphamide was improved by oral doses of *Lactobacillus johonsoni* and *Enterococcus hirae*. The outcome of a combinatorial approach of CpG oligodeoxynucleotides and an anti-interleukin-10 receptor (IL-10R) antibody in quelling tumors via TNF was augmented by *Alistipes shahii*. Gut inflammation, a side effect of immunotherapy was lowered by oral applications of *Bacterioides fragilis* and *Burkholderia cepacia*.

Interestingly, *Lactobacillus* is suggested to handle diarrhea induced by chemotherapy according to the European Society of Medical Oncology (ESMO) Clinical Practice Guidelines for Gastrointestinal Mucositis and the Multinational Association of Supportive Care in Cancer and International Society of Oral Oncology (MASCC/ISOO) (Fulbright et al, 2017; Vivarelli et al, 2019).

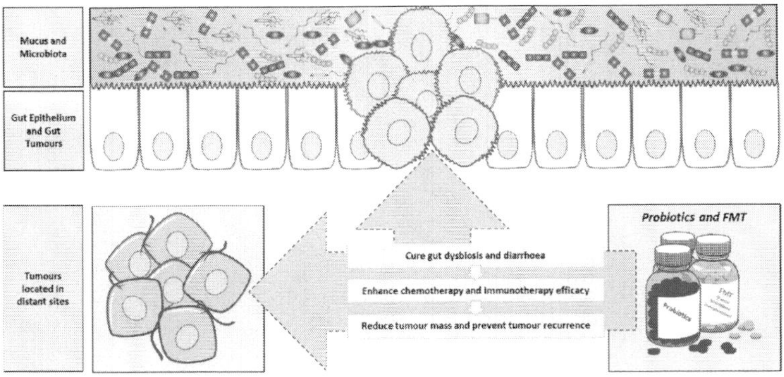

Figure 15: Above outlines the management of gut dysbiosis to manage cancer (Vivarelli et al, 2019: *Licensee MDPI, Basel, Switzerland. This article is an open-access article distributed under the terms and conditions of the Creative Commons Attribution (CC BY) license (http://creativecommons.org/licenses/by/4.0/*).

Continued research can continue to cast light on the complex and delicate associations of the gut microbes with various aspects of health. This is the time to truly understand the following quote:

"*Each patient carries his own doctor inside him.*"

— Norman Cousins, "Anatomy of an Illness"

4.5 REFERENCES

1. Mills KT, Bundy JD, Kelly TN, Reed JE, Kearney PM, Reynolds K, Chen J, He J. Global Disparities of Hypertension Prevalence and Control: A Systematic Analysis of Population-Based Studies From 90 Countries. Circulation. 2016 Aug 9;134(6):441-50. doi: 10.1161/CIRCULATIONAHA.115.018912. PMID: 27502908; PMCID: PMC4979614.
2. Mills, K.T., Stefanescu, A. & He, J. The global epidemiology of hypertension. *Nat Rev Nephrol* 16, 223–237 (2020). https://doi.org/10.1038/s41581-019-0244-2
3. Silveira-Nunes G, Durso DF, Alves de Oliveira LR Jr, Cunha EHM, Maioli TU, Vieira AT, Speziali E, Corrêa-Oliveira R, Martins-Filho OA, Teixeira-Carvalho A, Franceschi C, Rampelli S, Turroni S, Brigidi P and Faria AMC (2020) Hypertension Is Associated With Intestinal Microbiota Dysbiosis and Inflammation in a Brazilian Population. *Front. Pharmacol.* 11:258. doi: 10.3389/fphar.2020.00258
4. Yan Q, Gu Y, Li X, Yang W, Jia L, Chen C, Han X, Huang Y, Zhao L, Li P, Fang Z, Zhou J, Guan X, Ding Y, Wang S, Khan M, Xin Y, Li S and Ma Y (2017) Alterations of the Gut Microbiome in Hypertension. *Front. Cell. Infect. Microbiol.* 7:381. doi: 10.3389/fcimb.2017.00381
5. Li, J., Zhao, F., Wang, Y. *et al.* Gut microbiota dysbiosis contributes to the development of hypertension. *Microbiome* 5, 14 (2017). https://doi.org/10.1186/s40168-016-0222-x
6. Arima H, Barzi F, Chalmers J. Mortality patterns in hypertension. J Hypertens. 2011 Dec;29 Suppl 1:S3-7. doi: 10.1097/01.hjh.0000410246.59221.b1. PMID: 22157565.
7. Pouya Saeedi,Inga Petersohn,Paraskevi Salpea,Belma Malanda,Suvi Karuranga,Nigel Unwin,Stephen Colagiuri,Leonor Guariguata,Ayesha A. Motala,Katherine Ogurtsova,Jonathan E. Shaw,Dominic Bright,Rhys Williams. Global and regional diabetes prevalence estimates for 2019 and projections for 2030 and 2045: Results from the International Diabetes Federation Diabetes Atlas, 9th edition. Diabetes Research and Clinical Practice (2019); 157, 107843. DOI:https://doi.org/10.1016/j.diabres.2019.107843
8. Li, Q., Chang, Y., Zhang, K. *et al.* Implication of the gut microbiome composition of type 2 diabetic patients from northern China. *Sci Rep* 10, 5450 (2020). https://doi.org/10.1038/s41598-020-62224-3

9. Larsen N, Vogensen FK, van den Berg FWJ, Nielsen DS, Andreasen AS, Pedersen BK, et al. (2010) Gut Microbiota in Human Adults with Type 2 Diabetes Differs from Non-Diabetic Adults. PLoS ONE 5(2): e9085. https://doi.org/10.1371/journal.pone.0009085
10. Akshay H. Gaike, Dhiraj Paul, Shrikant Bhute, Dhiraj P. Dhotre, Pranav Pande, Smitha Upadhyaya, Yugandhar Reddy, Ramya Sampath, Debjani Ghosh, D. Chandraprabha, Jhankar Acharya, Gautam Banerjee, Vilas P. Sinkar, Saroj S. Ghaskadbi, Yogesh S. Shouche. mSystems Mar 2020, 5 (2) e00578-19; DOI: 10.1128/mSystems.00578-19
11. Lin HV, Frassetto A, Kowalik Jr EJ, Nawrocki AR, Lu MM, Kosinski JR, et al. (2012) Butyrate and Propionate Protect against Diet-Induced Obesity and Regulate Gut Hormones via Free Fatty Acid Receptor 3-Independent Mechanisms. PLoS ONE 7(4): e35240. https://doi.org/10.1371/journal.pone.0035240
12. Alessandra Puddu, Roberta Sanguineti, Fabrizio Montecucco, Giorgio Luciano Viviani, "Evidence for the Gut Microbiota Short-Chain Fatty Acids as Key Pathophysiological Molecules Improving Diabetes", Mediators of Inflammation, vol. 2014, Article ID 162021, 9 pages, 2014. https://doi.org/10.1155/2014/162021
13. Mattace Raso G, Simeoli R, Russo R, Iacono A, Santoro A, Paciello O, et al. (2013) Effects of Sodium Butyrate and Its Synthetic Amide Derivative on Liver Inflammation and Glucose Tolerance in an Animal Model of Steatosis Induced by High Fat Diet. PLoS ONE 8(7): e68626. https://doi.org/10.1371/journal.pone.0068626
14. Zhang X, Zhao Y, Zhang M, Pang X, Xu J, Kang C, et al. (2012) Structural Changes of Gut Microbiota during Berberine-Mediated Prevention of Obesity and Insulin Resistance in High-Fat Diet-Fed Rats. PLoS ONE 7(8): e42529. https://doi.org/10.1371/journal.pone.0042529
15. Boxun Zhang, Rensong Yue, Yuan Chen, Maoyi Yang, Xiaoying Huang, Jiacheng Shui, Yuliang Peng, Jiawei Chin, "Gut Microbiota, a Potential New Target for Chinese Herbal Medicines in Treating Diabetes Mellitus", Evidence-Based Complementary and Alternative Medicine, vol. 2019, Article ID 2634898, 11 pages, 2019. https://doi.org/10.1155/2019/2634898
16. Witchel, S. F., Oberfield, S. E., & Peña, A. S. (2019). Polycystic Ovary Syndrome: Pathophysiology, Presentation, and Treatment With Emphasis on Adolescent Girls. *Journal of the Endocrine Society*, *3*(8), 1545–1573. https://doi.org/10.1210/js.2019-00078

17. Lindheim L, Bashir M, Münzker J, Trummer C, Zachhuber V, Leber B, et al. (2017) Alterations in Gut Microbiome Composition and Barrier Function Are Associated with Reproductive and Metabolic Defects in Women with Polycystic Ovary Syndrome (PCOS): A Pilot Study. PLoS ONE 12(1): e0168390. doi:10.1371/journal.pone.0168390
18. Guo Y, Qi Y, Yang X, Zhao L, Wen S, Liu Y, et al. (2016) Association between Polycystic Ovary Syndrome and Gut Microbiota. PLoS ONE 11(4): e0153196. https://doi.org/10.1371/journal.pone.0153196
19. Overall, apart from the ability of specific microbes in the gut leading to carcinogenesis to intestinal commensals serving as adjuvant therapies in anti-cancer approaches, the role of gut microbiota in cancer and its therapy is indeed exciting!
20. Vivarelli, S., Salemi, R., Candido, S., Falzone, L., Santagati, M., Stefani, S., Torino, F., Banna, G. L., Tonini, G., & Libra, M. (2019). Gut Microbiota and Cancer: From Pathogenesis to Therapy. *Cancers*, *11*(1), 38. https://doi.org/10.3390/cancers11010038
21. Fulbright LE, Ellermann M, Arthur JC (2017) The microbiome and the hallmarks of cancer. PLoS Pathog 13(9): e1006480. https://doi.org/10.1371/journal.ppat.1006480

Chapter 5

GENETIC ANALYSIS OF GUT MICROBIOTA

"Treat our microbiome like a garden – we need to nourish the soil (intestines) for healthy plants (bacteria), while minimising weeds (disease-causing microbes.)"

- Tim Spector, a leading researcher in gut health

Dr Prerna Pandey[1]

[1]Dr Prerna Pandey, a biotechnologist with several years of wet lab research experience, has worked as a senior research fellow at International Center for Genetic Engineering and Biotechnology, New Delhi. Her field of research in PhD included isolation and molecular characterization of Geminiviruses, genome sequencing, gene annotation, as well as gene silencing using the RNA interference technology. She has published papers in peer-reviewed journals in the field and has a submitted a number of annotated Geminiviral genome sequences in the GenBank, including two novel ones. Prerna has also completed her editing and proof-reading courses from SfEP, UK and now works as a freelance scientific editor, writer and a subject matter expert. She has edited and co-authored several books on a range of topics in Life-Sciences. When Prerna is not busy with her assignments, she loves visiting historical places.

5.1 INTRODUCTION

The human intestines are home to a vast number of bacteria, archaea, microbial eukaryotes and viruses, collectively known as the gut microbiome1. The gut microbiome of healthy individuals is dominated by bacterial species.

A wide diversity of microbial species colonizes the human body, and provide considerable benefits through a range of different functions (Peter et al.,2019, Sekirov et al.;2010). Gut microbiome represents a vast number of bacteria, archaea, microbial eukaryotes and viruses which resides in the human intestine (Sender et al., 2016). In healthy individuals, bacterial species dominate the gut microbiome. These microbes produce metabolites that serve as energy sources for cell metabolism, promote the development and the functionality of the immune system, and prevent colonization by pathogenic microorganisms (Honda and Littman, 2012). Since birth, the gut microbiome is environmentally acquired (Costello et al., 2012; Walter and Ley, 2011), hence environmental factor that interacts with host genetics to shape phenotype, also as a genetically determined attribute that's shaped by, and interacts with, the host (Bevins and Salzman, 2011; Spor et al., 2011). Although the microbiome can be modified for therapeutic applications (Borody and Khoruts, 2012; Hamilton et al., 2013), it comprises an attractive target for manipulation. Once the interactions between host genetics and the microbiome are understood, its manipulation could be optimized for a given host genome to reduce disease risk (Julia et al.,2014).

Host genetics define the chemical and physical landscape inhabited by the gut microbiome, including nutrient availability and the threshold of activity for the immune system (Ley et al.,2008). As for instance, the composition of the gut microbiome is regulated by the host through the secretion of immunoglobulin A (IgA), antimicrobial peptides and microRNAs (Lehrer et al.,2012; Cullen et al.,2015). Thus, gut microbial communities are composed of specialists that have evolved to occupy ecological niches specific to the gut (Andrew et al.,2017), as well as the ability to metabolize specific molecules available from the host (Levin et al.,2017) or to evade host defences (Suzuki et al.,2004; Levin et al.,2017).

It is crucial to reveal the content, variety and functioning of the microbial intestine network as a way to recognize approximately its effect on human health. With the assist of 16S ribosomal RNA gene (rRNA) (Zoetendal et al.,1998) the 2 bacterial divisions, the Bacteroidetes and the Firmicutes, the 2 bacterial divisions have been revealed, which incorporate of just about 90% of the recognised phylogenetic classes and dominate the distal intestine microbiota (Eckburg et al.,2005). Studies additionally confirmed huge variety of the intestine microbiome among wholesome individuals (Turnbaugh et al.,2009). Although this distinction is mainly marked amongst infants (Palmer et al.2007), the intestine microbiome converges to greater comparable phyla in later life.

In this chapter we are going to discuss about the genetic significance of gut microbiota, how its gene catalogue is been established by using various techniques that can used to characterize the gut microbiota, once they are often applied to human studies and their relative advantages and limitations.

5.2 GUT BACTERIA AND ITS GENETIC SIGNIFICANCE

The numerous commensal bacterial groups in the gastrointestinal (GI) tract of people and animals offer essential capabilities which includes regulating development, growing the host's digestion capabilities, and stopping pathogen colonization (Round and Mazmanian,2009; Maslowski and Mackay,2011). Animal intestine microbial groups are received and formed dynamically after birth, and every now and then even earlier than birth; they may be complicated structures that offer health-applicable capabilities (Matamoros et al.,2013; Neu,2016). Genetic coevolution among hosts and intestine microbes has led to particular dietary symbiosis consisting of short-chain fatty acid manufacturing and diet synthesis (Morrison and Preston ,2016; Leblanc et al.,2013), which suggests the lifestyles of heritable microbiota. In a few host dietary advantages given with the aid of using intestine micro organism are so robust that a few microbe taxa look like preferentially selected, both through direct contact or through genetic mediated impact with the aid of using the parental genome. Due to the critical roles of heritable bacterial taxa, key modulators consisting of host genetics may also decide host–microbe specificity to keep microbial groups withinside the GI tract; bacterial colonization and filtering arise in reaction to non-stop ecological interactions. Exploring how host genotypes modulate intestine micro organism composition will deepen our information for setting up new goals to decrease intestinal issues and metabolic syndromes with the aid of using keeping homeostasis in the GI tract (Piexin et al.,2020).

Although the volume upto which host genetics form the intestine microbiota stays uncertain, a developing variety of research throughout animal species (e.g., human, mouse, chicken, cattle, and swine) have validated the position of host genetics withinside the intestine microbiota composition (Goodrich et al.,2014; Zhang et al.,2010). With latest superior technology, in particular genome-wide association study (GWAS), more than one institutions are diagnosed among unmarried nucleotide polymorphisms (SNPs) genotypes, normally positioned in genes associated with host metabolic syndrome and immune disease, and abundance of commensal bacteria (Bonder et al.,2016; Wang et al.,2016). However, currently the statistical importance of institutions among host SNPs and man or woman bacterial taxa has been challenged (Wen et al.,2019), and further research document that environmental elements dominate host genetics in shaping the intestine microbiota (Rothschild et al.,2018; Carmody et al.,2015). The discrepancies in decoding the position of host genetics and environmental elements will be way to populace variation, genetic distance, age, and environmental conditions. The interconnection among environmental elements and host genetics itself may also masks host genetic consequences in shaping the intestine microbiota.

The intestine microbiome is environmentally received from birth (Costello et al., 2012; Walter and Ley, 2011), consequently it'll feature as an environmental thing that interacts with host genetics to form phenotype, in addition to a genetically decided characteristic this is formed by, and interacts with, the host (Bevins and Salzman, 2011; Tims et al., 2011). Because the microbiome may be changed for healing applications (Borody and Khoruts, 2012; Hamilton et al., 2013), it constitutes an appealing goal for manipulation. Once the interactions among host genetics and consequently the microbiome are understood, its manipulation is probably optimized for a given host genome to cut back disorder risk.

The microbiome of human manifest both intra- and inter-individual variability (Lax et al., 2014; Gupta et al., 2017). Studies on twins have shown that the microbiota of identical twins are more similar compared thereto of their siblings. In comparison with unrelated individuals, siblings have a more similar microbiota (Goodrich et al., 2014). Similar to genetic variation, the human microbiome plays an important role in health and disease (Blekhman et al., 2015; Gilbert et al., 2018). The microbiome has been related to variants in host genes involved in immunity and metabolism (Blekhman et al., 2015). The human system features a complex bidirectional relationship with the microbiome. It has been shown that the microbiome is related to the variability of the immune responses and these responses can also be involved in modifying the microbiome itself (Belkaid and Hand, 2014; Takiishi et al., 2017). Additionally, human genetic polymorphisms at various loci are hypothesized to interact with each other and with an individual's microbiome to impact disease (Blekhman et al., 2015; Hall et al., 2017). In particular, mutations in host genes can influence its interaction with the compositional and functional diversity of the microbiome, potentially modulating an individual's susceptibility to disease (Sandoval-Motta et al., 2017). In healthy individuals, the microbiome composition is balanced (Rajiliæ-Stojanoviæ et al., 2009; Lloyd-Price et al., 2016), and imbalance is now known to be associated with clinical conditions being like inflammatory bowel disease and diabetes (van Tongeren et al., 2005; Upadhyaya and Banerjee, 2015). Besides host genetics other factors are related to microbial community composition including diet and antibiotic consumption. In many studies it have have reported that both long- and short-term diet can influence the microbiome composition (Wu et al., 2011; Conlon and Bird, 2014). For example, high-carbohydrate diets are related to prevalence of Prevotella, while Bacteroides are related to high-fat and high-protein diets (Singh et al., 2017). Moreover, consumption of antibiotics may shift the microbiome composition to a temporally quasi-stable state. This state are often either capable of reverting back to the initial state, or to an alternate irreversible post-antibiotic dysbiosis state (Lozupone et al., 2012). This dysbiosis state is marked by a loss of taxonomic and functional diversity, which may transfer the host's metabolic capacity and reduce the colonization resistance against invading pathogens (Langdon et al., 2016; Lange et al.,2016).

Effect of the host genetic at the microbiome comes ordinarily from research taking a focused technique. In case, the concordance fee for carriage of the methanogen *Methanobrevibacter smithii* is better for monozygotic (MZ) than dizygotic (DZ) dual pairs (Hansen et al., 2011), and research evaluating microbiotas among human topics distinguishable at particular genetic loci have proven gene-microbiota interactions (Frank et al., 2011; Khachatryan et al., 2008; Rausch et al., 2011). A extra complete technique to this query has related genetic loci with abundances of intestine microorganism in mice (Benson et al., 2010; McKnite et al., 2012), however in humans, a fashionable technique (e.g., the use of twins) has failed to expose great genotype results on microbiome diversity (Yatsunenko et al., 2012). Thus, heritable constituent of the human intestine microbiome continues to be to be recognized the use of an independent technique.

5.3 FACTORS AFFECTING HUMAN MICROBIOME COMPOSITION

Gut microbiota has crucial features withinside the body, and imbalances withinside the composition and variety of these microbiota can purpose numerous illnesses. The host promotes acceptable microbiota through freeing particular elements, which includes microRNAs, and nonspecific elements, which includes antimicrobial peptides, mucus and immunoglobulin A that inspire the increase of particular forms of microorganism and inhibit the increase of others. There are many elements which can alternate intestine microbiota like diet, antibiotics, and age, and plenty of research have proven the connection among problems of the microbiota and numerous illnesses and stated a few methods to modulate that balance (Hasan and Yang,2019).

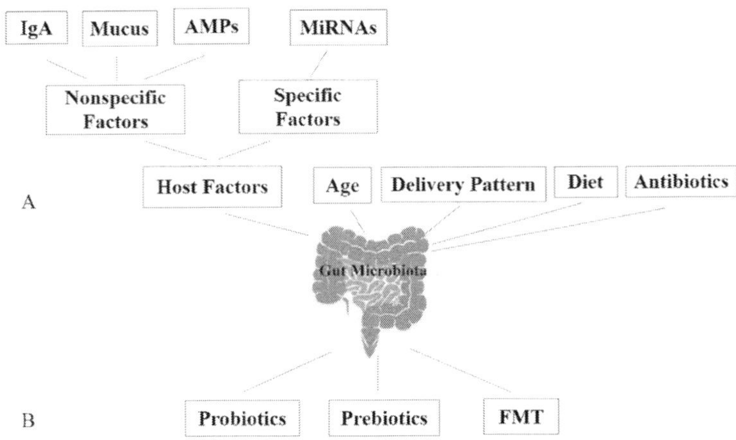

Figure 1: Factors affecting gut microbiota and ways to modulate it. (A) Factors affecting gut microbiota. (B) Ways to modulate gut microbiota. AMPs, antimicrobial peptides; IgA, immunoglobulin A; miRNA, microRNA; FMT, fecal microbiota transplantation. Copyright: © *2019 Hasan and Yang.* ***Licence:*** *This is an open access article distributed under the terms of the Creative Commons Attribution License, which permits unrestricted use, distribution, reproduction and adaptation in any medium and for any purpose provided that it is properly attributed.*

To cultivate their own gut microbiota hosts use factors which are specific as well as non-specific. The host approves only those type of microbes that can colonize its intestines for its good and removes other microbes which are harmful from the body.

Nonspecific Host Factors

The host selects its own gut microbiota by generating several molecular signals that dominate the structure of the surfaces colonized by microbiota and so influences its composition. These molecules are produced with the aid of intestinal epithelial cells (IEC) and include mucus, AMPs, and immunoglobulin A (IgA), that can encourage the growth of some microbial species and inhibit that of others. Mucus performs a crucial position in retaining the microbes some distance from IEC withinside the large intestine, it is constituted of two layers: the inner layer which does not contain any microorganisms whereas the outer layer contains soluble mucins, which can be adorned with the aid of using O-glycans, providing a nutrient source and binding site for gut microbiota (Artis et al., 2004). The essential position is performed with the aid of using mucus and mucin O-glycans in shaping the intestine microbiota and choosing the maximum suitable microbial species for host health (Arpaia et al., 2013). Gut microbiota utilise mucin which depend upon glycoside hydrolases and polysaccharide lyases which can be encoded with the aid of using their genes (Tailford et al., 2015). Some species of gut microbiota have capacity to degrade complicated carbohydrates like *Bacteroides thetaiotaomicron*, whose genome consists of 260 glycoside hydrolases (Cockburn & Koropatkin, 2016). Since there isn't always enough mucus withinside the small intestine, AMPs play an essential role in shaping gut microbiota. The gut microbiota via its structural components and metabolites bring about AMP production by Paneth cell by a mechanism that is mediated by pattern recognition receptor (PRR) (Hooper, 2009), whereas PRRs are operated with the aid of using exclusive microbial components, such as flagella and lipopolysaccharide, in a system called microbe-associated molecular patterns (MAMP). PRR-MAMP performs essential roles in promoting the function of the mucus barrier and inducing the production of IgA, mucin glycoproteins, and AMPs (Carvalho et al., 2012), and AMPs concentration are at maximum in small intestinal crypts because of Paneth cells inhabit at this location. IEC secretes AMPs as the body's first line of defense in response to invaders and these proteins have a broad effect that directly kills bacteria, virus, yeast, fungi, and even most of the cancer cells. These proteins comprise of defensins, cathelicidins, Reg family proteins, and ribonucleases (Cash et al., 2006). It has been proven that all dominant species of gut microbiota can tolerate high concentrations of AMPs, and AMPs might play an important role when the host distinguishes commensal from pathogenic bacteria (Cullen et al., 2015). The most abundant Gram negative genus, Bacteroides, among the gut microbiota, had tolerance to AMPs (Cullen et al., 2015). The sensitivity to Amps increased after *Bacteroides* strains lost their ability to produce dephosphorylate lipid A (Raetz & Whitfield, 2002), suggesting that dephosphorylate lipid A gives the Bacteroides their ability to withstand the high concentrations of AMPs in the intestine.

Antibacterial lectins, that could shape hexameric pores in Gram-positive bacterial membranes and stop them from reaching the intestinal mucus layer (Mukherjee et al., 2014), are crucial antibacterial factors that make a contribution in shaping gut microbiota. Based on these secreting factors, IEC help in the growth of some bacterial species and inhibit that of others so the host control the shape and structure of its gut microbiota by modulating these secretors.

Furthermore, plasma cells which are already their in the intestinal mucosa produce secretory immunoglobulin A (SIgA) which can cover the bacteria and manage its numbers locally (Macpherson & Uhr, 2004). Further, SIgA plays a bit part in bacterial biofilm formation by means of binding to SIgA receptors on bacteria (Randal Bollinger et al., 2003). The gut microbiota presence might possibly a condition to activate dendritic cells, which induce plasma cells to produce IgA (Massacand et al., 2008). The IgA absence can cause increases in segmental filamentous bacteria, recommending that the induced secretory IgA production is a form of competition between different types of gut microbiota (Suzuki et al., 2004).

Specific Host Factors

The form and shape of gut microbiota can be evaluated with the aid of using miRNAs, these are small non-coding RNAs which are 18–23 nucleotides in length. miRNAs are generated inside the nucleus after which it gets transported to the cytoplasm to effect gene silencing by hybridizing with the 3′ untranslated location of the target gene and promote degradation of mRNA or inhibit translation (Djuranovic, Nahvi & Green, 2012). One miRNA can target more than one type of mRNAs (Kalla et al., 2014), and it has been proven that miRNAs exist outside the cells and circulate in bodily fluids (Weber et al., 2010).

Ahmed et al. (2009) and Link et al. (2012) evaluated RNA in human stool and indicated that miRNAs has ability to behave as markers for intestinal malignancy, whereas Liu et al. (2016) seemed into the gut miRNAs in intestinal contents and feces. All these three works showed their ability to affect the composition of the gut microbiota composition. The main sources of fecal miRNA are IEC and Hopx-positive cells. Ahmed et al. (2009) and Link et al. (2012) additionally related the connection among the deficiency of IEC miRNA and gut dysbiosis and also studied why wild-type fecal transplantation restore the gut microbiota, which likely imply that miRNAs were capable to alter the gut microbiome. Various miRNAs can then enter the gut bacterial cells and can manipulate their growth and gene expression. Liu et al. (2016) mounted that hsa miRNA-515-5P can stimulate *Fusobacterium nucleatum* growth and that hsa miRNA-1226-5p also additionally stimulates E. coli growth through culturing the strains with synthesized in vitro mimics miRNA which affected E. coli in vivo also when given to mice orally. Hence, miRNA was found to have specific effects on gut bacterial growth.

In addition, Liu et al. (2016) demonstrated that miRNA can enter the bacterial cell and settle near bacteria DNA in the nucleus, and then, they studied the effect of synthetic miRNAs on bacterial gene expression and growth. The presence of gut microbiota has an impact on miRNA expression in the intestine. Dalmasso et al. (2011) discovered that nine miRNAs which are present in the colon and ileum had different levels of expression in the germ-free mice which are made to colonized with pathogen-free mice microbiota. Xue et al. (2011) discovered that microbiota can have an effect on expression of miR-10a, which targets interleukin-12/ interleukin-23p40B, which is also by itself responsible for innate immune response of the host toward the gut microbiota, suggesting that the gut microbiota can manipulate host innate immune responses by regulating of miRNA expression. The gut microbiota may be shaped by administration of fecal miRNAs. The microbiota profile of the recipient turns into just like that of the donor after transplantation of fecal RNA, as proven in the case of transfer from healthy mouse donors to IEC-deficient mice (Liu et al., 2016). This suggests an important potential of application, in the future, diseases related to changes in gut microbiota may be treated using synthetic specific miRNA.

The gut microbiota is established early in life stage but can later be altered by various factors that affect its development and diversity.

Age and Delivery Pattern

The method of microbic gut formation starts in utero by microbiota within the humor and placenta (Collado et al., 2016). Studies have aforementioned that there are microorganism and also their products, like DNA, in stool (Nagpal et al., 2017), amniotic fluid (Collado et al., 2016; DiGiulio et al.,

2008), and the placenta (Collado et al., 2016; Friedrich, 2013). Studies at the pregnant mice have shown that oral administration of tagged *Enterococcus fecium* resulted staring isolated it from the newborn stool samples (Jime ́nez et al., 2008), this result admit with the proof that maternal microbes is also transferred to the humor (Collado et al., 2016; Jime ́nez et al., 2008) and also the placenta (Goldenberg, Hauth & Andrews, 2000). when birth, the mode of delivery associate with the effect on the early-life development of the gut microbiota. The delivery of newborns through vagina have primary gut microbiota usually dominated by *Lactobacillus* and *Prevotella* derived from the mother's vaginal microbiota, whereas those born via cesarian section derive their gut microbiota from the skin, leading to dominance of *Streptococcus*, *Corynebacterium*, and *Propionibacterium* (Dominguez-Bello et al., 2010). These after birth microbiota evolve over time to become a lot of numerous and relatively stable. At the age of three years, they become mostly like adult's gut microbiota (Yatsunenko et al., 2012).

Diet

After birth, the first effect on the gut microbiota is the infant diet (breast or formula milk). The composition of the milk have an effect on shaping the early gut microbiota (Groer et al., 2014). In nursing or breastfeeding infants, the species that rule the gut microbiota are Lactobacillus and Bifidobacterium; breast milk contains oligosaccharides which can be broken down with ease by these species, leading to a rise in short-chain fatty acids, that directs the immune system to increase the expression of immunoglobulin G (Ouwehand, Isolauri & Salminen, 2002). Those infants who are raised on formula, the dominant species are *Enterococcus*, *Enterobacteria*, *Bacteroides*, *Clostiridia*, and *Streptococcus* (Yoshioka, Iseki & Fujita, 1983). The primary microbiota acquired during infancy may play a crucial role in initial immunity during the growth of babies, that's why, the composition of primary microbiota during this period is very important to protect babies from diseases related to poor immunity (Groer et al., 2014). Many studies have compared the gut microbiota population and mucosal immune response between breast and formula feeding and state that the breast feeding caused more stable population and good mucosal immune response (Bezirtzoglou, Tsiotsias & Welling, 2011).

It seems interesting that, human milk microbiota play vital roles in immunological activities namely increasing the number of plasma cells in the intestinal environment of newborns that lead to production of IgA (Gross, 2007), stimulation of cytotoxic Th1 cell (M'Rabet et al., 2008), motivates specific cytokines that make a balanced microenvironment (Walker & Iyengar, 2014), growth of local and systemic immune system (Houghteling & Walker, 2015), and a few *Lactobacillus* strains which promote the production of Th1, cytokines and TNF-, regulate T cell, and stimulate NK cells, CD4+ T-cells and CD8+ T-cells.

After infancy, the gut microbiota continues its development, and therefore the diet remains key to process the shape, structure, and variety of the gut microbiota. Vegetarian diets are found to be related with healthy, various gut microbiota characterized by the domination of species which will metabolize insoluble carbohydrates, such as *Ruminococcus*, *Roseburia*, and *Eubacterium* (Walker et al., 2011), whereas a non-vegetarian diet (Western diet) has been related with a decreasing number of *Firmicutes* and an increase in *Bacteroides* (David et al., 2014). Diet can cause vital changes even over short periods (David et al., 2014). After consumption of a Western diet, the gut microbiota ferment amino acids, that lead to production of short-chain fatty acids as energy sources, and harmful compounds can be produced (Windey, De Preter & Verbeke, 2012). The vegetarian diet inhibits this, and promote carbohydrate fermentation as the main function of microbiota (Clarke et al., 2014).

Antibiotics

The use of antibiotics may be a two-edged weapon: it destroys each pathological and beneficial microbes indiscriminately, permitting loss of gut microbiota or the alleged dysbiosis and growth of unwanted microbes (Klingensmith & Coopersmith, 2016). Studies on experimental mice have indicated the antibiotics administration affected secondary bile acid and serotonin metabolism in the colon which results in delayed enteric motility by causing microbiota depletion (Ge et al., 2017). Antibiotics disrupt the competitive exclusion machinery, a basic property by which microbiota remove pathological microbes (Hehemann et al., 2010). This disruption promotes growth of alternative pathogens, such as *Clostridium difficile* (Ramnani et al., 2012). According to studies clindamycin (Jernberg et al., 2007), clarithromycin and metronidazole (Jakobsson et al., 2010) and ciproflaxin (Dethlefsen & Relman, 2010) affect the microbiota structure for a long time.

Clindamycin causes changes in microbiota that last 2 years with no recovery within the diversity of genus Bacteroides (Jernberg et al., 2007). Similarly, the utilization of clarithromycin in *Helicobacter pylori* treatment causes a downfall in the population of *Actinobacteria* (Jakobsson et al., 2010), whereas ciprofloxacin has been shown to cause a decrease in *Ruminococcus* (Dethlefsen et al., 2008). Vancomycin, that is taken into account as the most effective treatment for *C. difficile* infection (CDI), but, like other antibiotics, it causes changes in gut microbiota that lead to recurrent *C. difficile* infection (rCDI) (Zar et al., 2007) or encourage growth of pathological strains of *E. coli* (Lewis et al., 2015). Vancomycin treatment additionally also causes depletion of most gut microbiota, such as *Bacteroidetes*, and it is associated with increases in *Proteobacteria* species (Isaac et al., 2016), and decreases in *Bacteroidetes*, *Fuminococcus*, and *Faecalibacterium* (Vrieze et al., 2014). The precise impact of antibiotic administration on the gut microbiota and therefore the recovery time are depending on individual, like effect of the changes within the gut microbiota before treatment (Dethlefsen & Relman, 2010; Jernberg et al., 2007).

In several countries, the antibiotics are used for agriculture significantly for intensive farming of poultry and beef, and low doses of antibiotics are routinely given to livestock to extend their growth and weight (Blaser, 2016). Many studies on human and rodent have indicated to an obesogenic impact of antibiotics in humans even in low doses found in food (Blaser, 2016). Pesticides and other chemicals are frequently sprayed on foods and currently, the proof is lacking for their harm on gut health and the effects of organic food (Lee et al., 2017).

5.4 ESTABLISHMENT OF HUMAN GUT MICROBIAL GENE CATALOGUE

Determining the taxonomical and purposeful composition of a healthy gut microbiome across completely different populations is important for understanding its role in maintaining human health. Many large-scale, world-wide microbiome projects have discovered the variability within the gut microbial composition of healthy individuals because of the factors such as mode of delivery, age, geographical location, diet, and lifestyle (Arumugam et al.,2011). Most of the gut microbiome studies have determined microbial taxonomy and functional diversity using 16S rRNA marker gene-based and/or whole-genome shotgun (WGS) approaches to know the functional role of the gut microbiome. However, novel insights on the complex interaction among diet, gut microbes, and human health, together with the role of key microbial metabolites such as short chain fatty acids (SCFAs) and branched chain amino acids (BCAAs), derived from the microbial fermentation of dietary fibers, are commencing to emerge from recent gut metabolomics studies (Schwiertz et al.,2010). Moreover, the direct impact of microbial metabolome on human health is additionally changing into apparent from the recent studies that specialize in the "gut microbiome-host metabolism axis" (Pedersen et al.,2016). Therefore, an integrative approach using both metagenome and metabolome-based characterizations of the gut microbiome seems pragmatic for gaining deeper purposeful and mechanistic insights into the role of gut microbes on human health (Dhakan et al.,2019).

The large-scale studies carried out so far chiefly represent the gut microbiome of urban populations primarily from Europe, the United States, and other different "WEIRD" countries (i.e., the Western, Educated, Industrialized, Rich, and Democratic countries) (Gupta et al.,2017). Only recently, some studies have characterized the human microbiome from various ethnic populations and found remarkable compositional variations as compared to the microbiomes from other earlier studied populations (Hasler et al.,2018). India is the seventh largest country in the world and harbors the second largest population spread across multiple geographical locations with enormous diversity in ethnicity, lifestyles, and dietary habits. India is home to the majority of the world's vegetarian population but also has an almost equal representation of its population consuming animal-based diets. The highest prevalence of diabetes in the world is of the Indian population (Mohan et al.,2007). According to the World Health Organization estimates (WHO, 2011), 53% of deaths in India in the year 2008 were attributed to metabolic conditions such as diabetes and cardiovascular diseases, which are predicted to reach ~75% by 2030 (WHO,2011).

Some studies have investigated the gut microbiome of the Indian population. A recent study by Maji et al. has shown the functional association of human gut microbiome dysbiosis with tuberculosis through a time-course study of six tuberculosis patients in India (Maji et al.,2018). However, other gut microbiome studies were mainly limited by small cohort sizes and amplicon-based (16S rRNA gene) sequencing and analysis (Pulikkan et al.,2018). Thus, several large-scale efforts are needed to identify the Indian population specific microbiome biomarkers and to understand the impact of the gut microbiome on health and disease in the Indian population along with global comparisons.

5.5 TECHNIQUES FOR DETERMINING HUMAN MICROBIAL GENE

A symbiotic relationship is shown between the gut microbiota with its host. In the early 20th Century, a relationship between the gut microbiota and disease was first put forward. Although, until the 1990s, knowledge about gut microbiota was limited as bacteriological culture was the only technique available for characterization of its composition. So far only a fraction (estimated at <30%) of the gut microbiota has been cultured. Since the beginning of 1990s, advancement in culture-independent techniques have extended our knowledge with respect to the complexity of ecosystem. These techniques have explained the microbial diversity of the gut microbiota and disclosed that the alterations in the gut microbiota composition and function are linked with certain disease states, such as IBD and obesity. These techniques facilitate high throughput, are fast, identify organisms that are uncultured to date and enable enumeration of organisms that are present in the gut microbiota. Below we have discussed the techniques that can be used to characterize the gut microbiota when they are applied to human studies as well as their relative advantages and limitations (Fraher et al.,2012).

Table 1 | Techniques used to characterize the gut microbiota

Technique	Description	Advantages	Disadvantages
Culture	Isolation of bacteria on selective media	Cheap, semi-quantitative	Labour intensive, <30% of gut microbiota have been cultured to date
qPCR	Amplification and quantification of 16S rRNA. Reaction mixture contains a compound that fluoresces when it binds to double-stranded DNA	Phylogenetic identification, quantitative, fast	PCR bias, unable to identify unknown species
DGGE/TGGE	Gel separation of 16S rRNA amplicons using denaturant/ temperature	Fast, semi-quantitative, bands can be excised for further analysis	No phylogenetic identification, PCR bias
T-RFLP	Fluorescently labelled primers are amplified and then restriction enzymes are used to digest the 16S rRNA amplicon. Digested fragments separated by gel electrophoresis	Fast, semi-quantitative, cheap	No phylogenetic identification, PCR bias, low resolution
FISH	Fluorescently labelled oligonucleotide probes hybridize complementary target 16S rRNA sequences. When hybridization occurs, fluorescence can be enumerated using flow cytometry	Phylogenetic identification, semi-quantitative, no PCR bias	Dependent on probe sequences—unable to identify unknown species
DNA microarrays	Fluorescently labelled oligonucleotide probes hybridize with complementary nucleotide sequences. Fluorescence detected with a laser	Phylogenetic identification, semi-quantitative, fast	Cross hybridization, PCR bias, species present in low levels can be difficult to detect
Cloned 16S rRNA gene sequencing	Cloning of full-length 16S rRNA amplicon, Sanger sequencing and capillary electrophoresis	Phylogenetic identification, quantitative	PCR bias, laborious, expensive, cloning bias
Direct sequencing of 16S rRNA amplicons	Massive parallel sequencing of partial 16S rRNA amplicons for example, 454 Pyrosequencing® (Roche Diagnostics GMBH Ltd, Mannheim, Germany) (amplicon immobilized on beads, amplified by emulsion PCR, addition of luciferase results in a chemoluminescent signal)	Phylogenetic identification, quantitative, fast, identification of unknown bacteria	PCR bias, expensive, laborious
Microbiome shotgun sequencing	Massive parallel sequencing of the whole genome (e.g. 454 pyrosequencing® or Illumina®, San Diego, CA, USA)	Phylogenetic identification, quantitative	Expensive, analysis of data is computationally intense

Abbreviations: DGGE, denaturing gradient gel electrophoresis; FISH, fluorescence *in situ* hybridization; qPCR, quantitative PCR; TGGE, temperature gradient gel electrophoresis; T-RFLP, terminal restriction fragment length polymorphism.

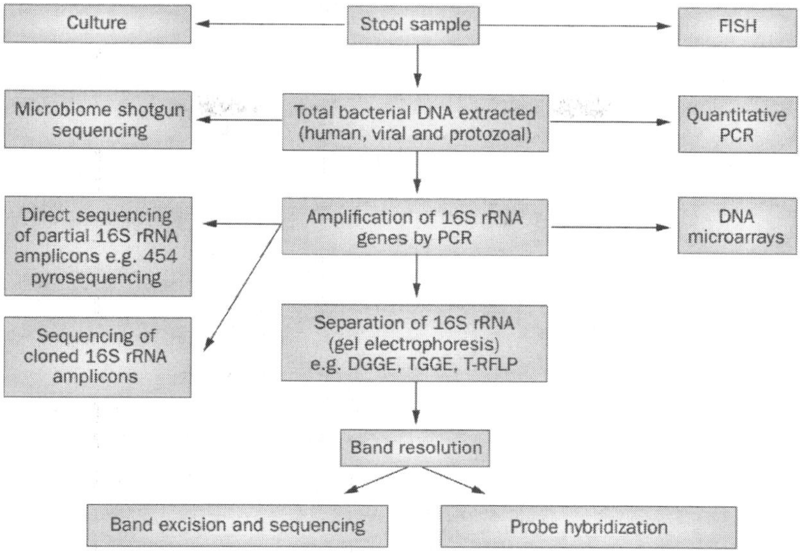

Figure 2: Overview of techniques used to characterize the gut microbiota (Fraher, M. H. *et al. Nat. Rev. Gastroenterol. Hepatol.2012;* 9, 312–322).

5.5.1 Cultures

Until 1990s, knowledge of the gut microbiota had been limited to culture-based techniques, an approach that was used since the early period of 20th century. Since then, advances were made in the phenotyping of isolates on the basis of their fermentation profiles and *in vitro* growth demands. Although bacterial identification by culture is cheap, it is labour intensive and the culture gives a limited view of the diversity of the gut microbiota as <30% of gut microbiota members have only been cultured to date (Eckburg et al.,2005). It is important to understand that uncultured organisms in the gut microbiota are not necessarily unculturable. They might, in fact, be culturable but permissive growth conditions for these organisms are yet to be developed or discovered.

In Parallel with the development of culture independent techniques, culture techniques became a lot more sophisticated through the utilization of, for example, gel microdroplets and microbial culture chips ('microPetri dishes') (Ingham et al.,2007). These techniques have high productivity and helps to culture previously uncultured organisms. Even though, some gut microbiota components have shown a symbiotic relationship with one another whereby such organisms rely on the metabolic activity of other organisms for the growth, thus having a limit on the utility of new pure-culture techniques (Stams,1994).

5.5.2 Quantitative Polymerase Chain Reaction (qPCR)

Studies on the gut microbiota with the help of qPCR alone have displayed changes concerning to the patient's age, antibiotic use and disease state (for example, Crohn's disease) (Mariat et al.,2009). Some molecular techniques which are used to characterize the gut microbiota (for example, DGGE or DNA microarrays) are semi-quantitative (see below) and are, therefore, not optimal for numerical assessment of the gut microbiota components (Rintilla et al.,2004). Once qPCR is employed together with these techniques it will offer a lot of elaborate data on the diversity and abundance of the gut microbiota (Ponnusamy et al.,2011). For example, Zwielehner *et al.* compared the gut microbiota in institutionalized aged patients with young healthy volunteers using DGGE and qPCR. The researchers demonstrated less bacterial abundance and total bacterial diversity within the institutionalized aged patients than in young healthy volunteers.

Advantages of qPCR include phylogenetic discrimination, speed and the inherent ability to design primers to target specific species; it is the most accurate culture independent measure of the total microbial load. Furthermore, qPCR kits are widely available commercially and can be used by laboratories that do not have access to the other techniques described. Disadvantages include the possibility of PCR bias unless extensive prior analysis has shown it not to occur for a given target– primer combination, and an inability to identify novel species (Sekirov et al.,2010). This technique can also be labour intensive and technically challenging (for example, primer design).

5.5.3 Denaturing Gradient Gel Electrophoresis and Temperature Gradient Gel Electrophoresis (DGGE/TGGE)

DNA fingerprinting is a community analysis tool that permits comparison of DNA fragments in a sample. DGGE disintegrates complex mixtures of 16S rRNA gene amplicons (for example, those derived from a stool sample), which are of similar length however have varying DNA sequences (Muyzer et al.,1993).

The 16S rRNA gene amplicon mixture is to be applied directly onto a polyacrylamide gel that has a linearly increasing gradient of denaturant (formamide and/or urea). On application of an electrophoretic current, the 16S rRNA gene amplicons migrate down the gel. Initially, the fragments tend to move according to their molecular weight alone however with increasing exposure to denaturant, the DNA strands start to denature. When they have almost fully denatured, migration halts. DNA sequence variation results in completely different melting temperatures; migration halts at different positions in the denaturing gradient, leading to separation of the segments along with 'band' formation. As sequence variation is found between species, separation of species is feasible. The gel is stained to permit band visualization, which in turn can be excised for sequencing or probe hybridization. This method permits separation of multiple 16S species from a mixed sample.

DGGE is principally used for comparative purposes; for example, to compare one disease to another or the healthy state (Noor et al.,2010). This technique is quick and multiple samples can be analysed at the same time (Muyzer,1999). This technique (along with TGGE, T-RFLP and FISH) is typically considered as quantitative because it provides visual impressions of band and/or species intensity and abundance. However, to be precise, these techniques are semi-quantitative at the best, because varying amplification dynamics create precise quantitative comparison inadvisable. Disadvantages embody PCR bias and the absence of direct phylogenetic identification unless sequencing or probe hybridization is done. TGGE has conjointly been employed in the study of the gut microbiota and it works in a similar manner to DGGE, however a linear temperature gradient is employed rather than a denaturing gradient gel (Osborn et al.,2000).

5.5.4 Terminal Restriction Fragment Length Polymorphism (T-RFLP)

T-RFLP is based on fragmentation of the 16S rRNA gene amplicons by restriction endonucleases with band visualization (Marsh,1999). T-RFLP is mainly used to assess microbial community diversity, for comparison purposes and also to assess variability in the gut microbiota (McCartney,2002). This technique is semi-quantitative, rapid and cheap. A limitation of T-RFLP is that it does not allow phylogenetic identification. However, this can be overcome by combining 16S rRNA clone library analysis with T-RFLP (Li et al.,2007).

5.5.5 Fluorescence In-Situ Hybridization (FISH)

Fluorescently labelled oligonucleotide probes that interbreed complementary target 16S rRNA sequences are being used in FISH. FISH can be carried out directly on the stool sample. The combination of FISH and flow cytometry has been shown to be a high-throughput method for analysis of the gut microbiota (Rigottier-Gois et al.,2003).

FISH is mainly used to compare the gut microbiota between disease states and health, often in the context of studying the differential abundance of a small number of organisms or families (Swidsinski et al.,2008). As mentioned on top of, the evident composition of the gut microbiota shall be altered with freezing. Rochet *et al.* examined the steadiness of the gut microbiota composition via *in situ* hybridization and flow cytometry at day 1 and months 1, 2, 4, 6, 8, 10 and 12 to quantify variations in the gut microbiota over storage time. Faecal suspensions were fixed overnight in 4% paraformaldehyde solution which then frozen at –70 °C within 12h of sample collection. This group demonstrated stability of the gut microbial composition for up to 8 months after sample collection. Storage of samples until further analysis gives flexibility to epidemiological studies. This technique is semiquantitative and fast which makes it more reliable for use. However, it is not possible to identify unknown species using this technique, but probe scan be designed to target specific phyla or species.

5.5.6 DNA Microarrays

DNA microarray technology (also known as gene array, phylogenetic microarray and DNA chip) is high throughput and provides phylogenetic identification of the gut microbiota (Palmer et al.,2007). Studies have shown that DNA microarrays can detect bacterial DNA present at a level of 0.00025% of the DNA sample (Paliy et al.,2009). Microarray chips targeting the gut microbiota have been developed and some are commercially available, including for example, the human intestinal tract chip (HITChip™; Agilent Technologies Inc., Santa Clara, CA, USA) and the PhyloChip™ (Affymetrix, Santa Clara, CA, USA) (DeAngelis et al., 2011).

DNA microarrays are primarily employed in studies to check the microbiota between completely different populations (Palmer et al., 2007). DNA microarray technique is quick, semi-quantitative and all species sample can be examined at the same time. Though this method is additionally qualitative, the accuracy of the qualitative information it provides has been debated (Cook and Sayler, 2003). The chance of cross-hybridization

(hybridization of multiple probes to single targets) makes it little difficult to use for analysis (Spiegeiman et al., 2005). It might be tough to discover species which are present at lesser amount in the gut, but detection is also possible if a reference sequence is available (Palmer et al.,2006).

5.5.7 Cloned 16S rRNA gene sequencing

70S ribosomes are distributed throughout the cytoplasm of a bacterial cell. They are composed of two subunits: 30S and 50S. The 50S subunit comprise of two RNA molecules: 5S and 23S. The 30S subunit (or small subunit) include one RNA molecule: 16S ribosomal rRNA (16S rRNA). One of the key role of 16S rRNA is to initiate and extend protein synthesis. As rRNA (5S, 16S and 23S) is extremely preserved between bacterial species, but also contains variable regions that yield a phylogenetic signal, which eventually act as a useful target for phylogenetic identification (bacterial identification). Among the three bacterial rRNA genes, the 16S rRNA gene provides the foremost tractable combination of preserved sites for PCR primers and variable regions that act as evolutionary chronometers, and it is, consequently, utilised in preference to 5S or 23S rRNA genes for phylogenetic identification (Rajendhran and Gunasekaran,2011). Most (but not all) culture-independent techniques being applied for the analysis of the gut microbiota are hence based on analysis of the 16S rRNA gene.

5.5.8 Direct Sequencing of 16S rRNA Amplicons

Massive parallel sequencing of limited 16S rRNA gene amplicons uses beads (454 Pyrosequencing®), slides (Illumina®) and solid surfaces (SOLiD™). It is described as massively parallel sequencing as 'massive' numbers of DNA templates can be sequenced in parallel, that is, at the same time and in the same reaction set-up. Approximately 500 million bases can be sequenced by Pyrosequencing with 99% or better accuracy, in one go (Voelkerding et al.,2009). It leads to an approximate 2,000-fold increase in throughput over Sanger sequencing. Since shorter sequences can be read (approximately one half of the read lengths generate in Sanger sequencing), many more sequences can be read and therefore bacteria that are in low abundance can also be detected (Rogers et al.,2005). 20,000–40,000 454 titanium reads of around 400 nucleotides per subject are routinely generated in microbiota studies, in line with outcomes from phylotype discovery studies (Claesson et al.,2009). Additional advantages of this technique include phylogenetic identification; the detection and/or identification of unknown bacteria;

speed; and the ability to collect quantitative data. Massively parallel sequencing is moderately expensive—it costs an estimated US$5,000 to sequence a novel microbial genome to high standard and less than $500 to resequence a genome from a species for which a reference genome is available. The basic uses of 454 Pyrosequencing® are to compare the gut microbiota between different states—for example, lean versus obese—and also to detect the effect of antibiotics on the gut microbiota (Turnbaugh et al.,2009); however, many more applications are possible. Jeffery *et al.* used 454 Pyrosequencing® to compare the microbiota between patients with IBS and healthy individuals. They found that the gut microbiota of patients with IBS formed two clusters; those with abnormal microbiota (an increased Firmicutes: Bacteroidetes ratio) and those with normal microbiota (similar to healthy controls). Interestingly, those patients with IBS and normal microbiota demonstrated clinically significant depression and/or anxiety using the hospital anxiety and depression scale. The authors conclude that this finding requires confirmation and further analysis; however, this study highlights the potential of next generation sequencing in understanding the pathophysiology of not only IBS but other diseases too.

5.5.9 Microbiome shotgun sequencing

Microbiome shotgun sequencing therefore imply shotgun sequencing of whole community DNA; which is accomplished by massive parallel sequencing of the mixed DNA sample. Shotgun sequencing, as the name suggests, involves random fragmentation of DNA, sequencing of DNA fragments and reconstruction of overlapping sequences to assemble them into a continuous sequence. Since other techniques analyse the genetic diversity of the gut microbiota only but microbiome shotgun sequencing collects data and analyse the genetic diversity as well as functions of the gut microbiota—other techniques analyse the genetic diversity of the gut microbiota alone. This analysis on the genetic diversity and functions of the gut microbiota permits correlations to be made between healthy and diseased states (De Vos,2009). Restrictions of microbiome shotgun sequencing are that it is expensive and the fact that the analysis of the large amount of data generated is computationally intense and not nevertheless simply performed in nonspecialist laboratories.

5.5.10 Metagenomic Sequencing

Metagenomic sequencing represents a robust alternative to rRNA sequencing for examining complicated microbial communities (Von Mering etal.,2007).

Such studies have already created some 3 Gb of microbial sequence from faecal samples of 33 individuals from the United States or Japan as applied to the human gut (Gill et al.,2006).

Metagenomics of the human gastrointestinal tract was first described in 2006 when Breitbart *et al.* characterized approximately 1,200 viral genotypes from human faeces. Manichanh *et al.* were the first group to use the metagenomic approach to sequence cloned fragments from the gut microbiota. Pooled stool samples from healthy volunteers and patients with Crohn's disease were compared, and results revealed that the community complexity of Firmicutes was reduced in the faecal microbiota of patients with Crohn's disease. Gill *et al.* described humans as 'superorganisms' whose metabolism represents a combination of human and microbial functions. This group demonstrated that the gut microbiota is enriched with genes that are essential for host functions, for example vitamin biosynthesis. The HMP is an NIH-led initiative that aims to characterize the human microbiota at five major sites (nasal and oral cavities, gastrointestinal and urogenital tracts and skin) and analyse the role of these microbes in human health and disease (Peterson et al.,2009).

5.5.11 GENOME-WIDE ASSOCIATION STUDIES

Advances in genotyping made feasible, at a relatively low cost, the genotyping of hundreds of thousands (or even millions) of common variants across the human genome. Together these factors catapulted the genome-wide association study (GWAS) in humans (herein referred to as "host") population.

Genome-wide association study in host populations (hGWAS) has helped us to know many genetic variants which are related to several complicated human traits and diseases, novel biological mechanisms and drug targets for infectious and non-infectious diseases (Relling and Evans, 2015). The microbiome, which known to be assembly of bacteria, archea, fungi, protozoa, and viruses that harbor our body surface and their respective genome (Blum, 2017), has been shown to play a critical role in human health and disease. The success of hGWAS prospective has provided an optimistic viewpoint for an eventual implementation to the microbiome for detection that can be used for the welfare of human benefits. The microbiome genome-wide association study (mGWAS) targets to recognise the host's genetic polymorphisms which interconnect with its microbiome. Recently, mGWASs have identified and validated various heritable bacterial taxa, including the

Christensenellaceae and *Methanogens* families (Goodrich et al., 2017). Besides, mGWAS has linked host genotypes and identified pathways with inter-individual variability in microbiome composition in states of health and disease (Hall et al., 2017; Imhann et al., 2018). These findings verified the general view regarding microbiome playing a noteworthy role in a host's traits, disease susceptibility and resistance, and treatment response to various diseases.

Table 2: Examples illustrating partial major technological and large-scale collaborative projects (excluding data repositories) on host and microbiome genome-wide association studies (GWAS)

Year	Achievements
1970s	Development of DNA sequencing technologies (Sanger sequencing)
1990	-Development of improved sequencing methods: PCR, RFLP, PFE -Commencement of the Human Genome Project -Onset of Biobanks
2001-2005	-Publication of 1st draft of human genome by HGP -Commencement of the International HapMap Project -Completion of HGP -Initiation of the Biobank Japan Project -Unveiling of the first NGS technology- Roche 454Life science GS 20 system -Completion of Phase I of the International HapMap Project -The Wellcome Trust Case control Consortium -2005-2009: Development of SNP genotyping technologies
2006-2009	Initiation of UK Biobank -Completion of Phase II of IHMP and its extension to Phase III -Sequencing of 1st human diploid gene by Sanger Sequencing Insights of human genetic variation begin to emerge with publications on human genetic variations -Initiation of Human Microbiome project (HMP) -initiation of Metagenomics of the Human Intestinal Tract (MetaHIT) -Initiation of the International Human Microbiome Consortium (IHMP) -Sequencing of 2nd diploid gene byNGS technology -Initiation of 1000 genome project -Introduction of Heliscope single molecule sequencer -Completion of Phase III of the International HapMap Project -Establishment of The US National Cancer Human Biobank
2012-onwards	-Initiation of UK 100,000 Genome Project -Initiation of the Integrative Human Microbiome Project Consortium (iHMP) -Initiation of African Genome Variation Project - Completion of 1000 genome project -Initiation of Australian 100,000 Genome Project - Initiation of the Integrative Human Microbiome Project Consortium (iHMP)

5.6 IMPACT OF GUT MICROBIOTA ON HUMAN HEALTH

Generally, gut bacteria and the host live in a mutualistic manner. On one hand, they will offer essential nutrients, synthesize vitamin K, support in the digestion of cellulose, and promote angiogenesis along with enteric nerve function (Hill and Artis,2010). Bacteroidetes and Firmicutes are the main bacteria in the metabolism of undigested food residues. They assist digestion of dietary fiber and polyphenols through a complex metabolic energy-harvesting mechanism that relies on cross-feeding and co-metabolism. In return, commensal bacteria cash in on the host's protective and nutrient-rich surrounding (Tsuji et al.,2008). Nevertheless, specialized gut bacteria tends to carry out reductive reactions such as methanogenesis, acetogenesis, nitrate and sulphate reduction (Tuohy et al.,2006). On the other hand, commensal bacteria and probiotics will promote barrier integrity, and stop antigens and pathogens from coming into the mucosal tissues (Ulluwishewa et al.,2011). In addition, commensal bacteria contribute to the host defence by regulating the homeostasis of the host's immune system (Macfarlane and Cummins,1999). However, gut bacteria can be potentially harmful once the gut ecosystem undergoes unusual changes. Dysbiosis of the gut bacteria communities in patients or animal models might cause allergy, inflammatory bowel disease (IBD), obesity, diabetes, and even cancer (Mueller and Macpherson,2006). The composition of the gut bacteria can depict the risk of diseases in each and every person (Akaza,2012). Understanding the relationship between gut bacteria and human health may be useful to find new probiotic treatments and novel methods to treat and manage a wide variety of human diseases (Zhang et al.,2015).

Gut bacteria have a crucial role in human health. It contributes to the host gut defence system and supports the gut to maintain all of its traditional functions, while its composition can be influenced by the host (Figure 3).

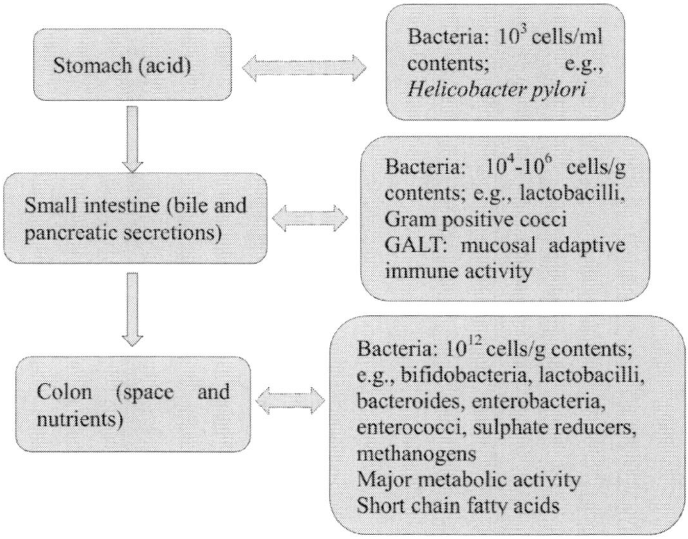

Figure 3: Reciprocal relationship between human gut bacteria and the host (Zhang et al.2015, doi:10.3390/ijms16047493; © *2015 by the authors; licensee MDPI, Basel, Switzerland. This article is an open access article distributed under the terms and conditions of the Creative Commons Attribution license (http://creativecommons.org/licenses/by/4.0/).*

Commensal bacteria contribute to the host gut defence system mainly by resisting the invasion of pathogenic bacteria and helping the development of the host immune system. Gut bacteria maintain resistance against the colonization of pathogenic bacteria by competing for nutrients and attachment sites on the mucosal surface in the colon, a phenomenon collectively known as "colonization resistance" (Stecher and Hardt,2008). The invasion of pathogenic bacteria is also prevented by commensal bacteria due to the reduction of the intestinal pH by the production of lactate and short-chain fatty acids (SCFAs) (Guarner and Malagelada,2003). Another way is by producing toxic or carcinogenic metabolites to inhibit the growth or kill potentially pathogenic bacteria, together with volatile fatty acids that can inhibit the colonization of pathogenic bacteria. For example, proteolytic fermentation in the distal colon could produce toxic, carcinogenic metabolites such as bacteriocins, ammonia, indoles, and phenols by gut bacteria (Beaud et al.,2005).

Not only do gut bacteria aid the host by contributing to the host gut defence system, they conjointly facilitate the gut to take care of traditional functions. Gut bacteria benefit the host in different ways, such as regulation of gut motility, manufacturing vitamins, transforming bile acid and steroids, metabolizing xenobiotic substances, absorbing minerals, and activating and eradicating toxins, genotoxins, and mutagens (Macfarlane et al.,2009). The proximal surrounding region of colon produces an excellent amount of short-chain organic acids, such as acetic, propionic, and butyric acids. Theses organic acids are energy sources for the colonic mucosa and peripheral body tissues, and they are metabolites of undigested complex carbohydrates by colonic bacteria fermentation. In return, theses organic acids have an effect on bacterial growth within the colon by affecting colonic water absorption and reducing fecal pH.

Generally, gut bacteria and the host live in a mutualistic manner. However, gut bacteria can be potentially harmful once the ecosystem undergoes unusual changes. Dysbiosis of the gut bacteria communities in patients or animal models may cause many diseases. Such as, antibiotic treatment and surgery leads to pseudomembranous colitis because of toxin production by *Clostridium difficile* as well as sepsis of *Escherichia coli*, *Enterococcus faecalis* and *Enterococcus faecium*, and intra-abdominal abscesses due to *Bacteroides fragilis* (Wilcox,2003). Imbalance in gut bacteria composition was associated with intestinal symptoms, such as bloating, abdominal pain, and diarrhoea. Uncultured phylotypes from *Clostridium* clusters XIVa and IV had statistically significant positive correlation along with bloating. *Ruminococcus callidus, Anaerotruncus colihominis* and *Lachnospira pectinoschiza* were on a higher side once bloating was recorded. The abundance of bifidobacteria may have a negative correlation with abdominal pain. Additionally, to bifidobacteria, one phylotype within *Ruminococcus lactaris*, *Clostridium* cluster IV, was significantly reduced in pain associated samples. Among the uncultured, positive correlations, potentially pathogenic phylotypes within uncultured *Clostridiales* II, *Anaerotruncus colihominis* and *Ruminococcus callidus*, were inflated over ten-fold when pain was recorded (Jalanka et al.,2011). Diarrhea was associated with a reduced number of members from the genus *Streptococcus*, particularly *S. alactolyticus* (Hermann et al.,2013). Besides, several different diseases were connected to gut bacteria, for example inflammatory bowel diseases, liver diseases, diabetes, cancers, chronic heart diseases, obesity, HIV, and autism.

5.7 SUMMARY

Taken along, the discoveries we have reviewed above suggest that for further advancing our knowledge of health and disease, we will be needing an improved characterization of the variability in the microbiota, a better understanding of how such variability can result in similar or different functional profiles, what cause such variation and more integrative studies that take into consideration the interaction between the microbiota, the host, and the environment to produce a specific phenotype and how host react to such factors. Most of the disease states and disorders are being found to correlate with the host microbiota (Virgin and Todd, 2011), including susceptibility to influenza (Ichinohe et al., 2011), retrovirus transmission (Kane et al., 2011), colon cancer (Kostic et al., 2011), autoimmune demyelination (Kostic et al.,2011), and even behavior (Heijtz et al., 2011). Large-scale sequencing projects such as the Human Microbiome Project (Peterson et al., 2009; Turnbaugh et al., 2007) and the Earth Microbiome Project (Gilbert et al., 2010) will eventually be critical for providing a unified and all-encompassing view to better understand the link between the microbiota and health. Machine learning approaches will be required in order to transform the enormous amount of data these projects generate so that it will be used in creating applicable and useful information for human well-being. The application of these methods will hopefully be more accurately predict disease states and appropriate therapies by making use of larger numbers of samples across populations and over time (Knights et al., 2011).

5.8 REFERENCES

1. Ahmed FE, Jeffries CD, Vos PW, Flake G, Nuovo GJ, Sinar DR, Naziri W, Marcuard SP. 2009.
2. Diagnostic microRNA markers for screening sporadic human colon cancer and active ulcerative colitis in stool and tissue. Cancer Genomics-Proteomics 6(5):281–295.
3. Arpaia N, Campbell C, Fan X, Dikiy S, Van Der Veeken J, deRoos P, Liu H, Cross JR, Pfeffer K, Coffer PJ, Rudensky AY. 2013. Metabolites produced by commensal bacteria promote peripheral regulatory T-cell generation. Nature 504(7480):451–455
4. Artis D, Wang ML, Keilbaugh SA, He W, Brenes M, Swain GP, Knight PA, Donaldson DD, Lazar MA, Miller HRP, Schad GA, Scott P, Wu GD. 2004. RELM/FIZZ2 is a goblet cell specific immune-effector molecule in the gastrointestinal tract. Proceedings of the National Academy of Sciences of the United States of America 101(37):13596–13600
5. Arumugam M, Raes J, Pelletier E, et al.2011. Enterotypes of the human gut microbiome. Nature ;473(7346):174–80.
6. Andrew Brantley Hall, Andrew C. Tolonen and Ramnik J. Xavier .2017. Human genetic variation and the gut microbiome in disease. Nat Rev Gene;18(11):690-699.
7. Beaud,D.;TailliezP.;Anba-Mondoloni;J.2005.Genetic characterization of the β-glucuronidase enzyme from a human intestinal bacterium, *Ruminococcus gnavus*. Microbiology, *151*, 2323–2330.
8. Belkaid, Y., and Hand, T. W. 2014. Role of the microbiota in immunity and inflammation. *Cell* 157, 121–141
9. Benson AK, Kelly SA, Legge R, Ma F, Low SJ, Kim J, Zhang M, Oh PL, Nehrenberg D, Hua K, et al.2010. Individuality in gut microbiota composition is a complex polygenic trait shaped by multiple environmental and host genetic factors. Proc Natl Acad Sci U S A.; 107:18933–18938.
10. Bevins CL, Salzman NH.2011. The potter's wheel: the host's role in sculpting its microbiota. *Cell* Mol Life Sci.;68:3675–3685.
11. Bezirtzoglou E, Tsiotsias A, Welling GW. 2011. Microbiota profile in feces of breast-and formulafed newborns by using fluorescence in situ hybridization (FISH). Anaerobe 17(6):478–482

12. Blaser MJ. 2016. Antibiotic use and its consequences for the normal microbiome. Science 352(6285):544–545
13. Blekhman, R., Goodrich, J. K., Huang, K., Sun, Q., Bukowski, R., Bell, J. T., et al. 2015. Host genetic variation impacts microbiome composition across human body sites. Genome Biol. 16, 1–12.
14. Blum, H. E.2017. The human microbiome. Adv. Med. Sci. 62, 414–420.
15. Bonder MJ, Kurilshikov A, Tigchelaar EF, Mujagic Z, Imhann F, Vila AV, et al.2016. The effect of host genetics on the gut microbiome. Nat Genet.;48:1407–12.
16. Borody TJ, Khoruts A.2012. Fecal microbiota transplantation and emerging applications. Nat Rev Gastroenterol Hepatol;9:88–96.
17. Breitbart, M. et al.2003 Metagenomic analyses of an uncultured viral community from human feces. J. Bacteriol. 185, 6220–6223.
18. Carmody RN, Gerber GK, Luevano JM Jr., Gatti DM, Somes L, Svenson KL, et al.2015. Diet dominates host genotype in shaping the murine gut microbiota. Cell Host Microbe;17:72–84.
19. Carvalho FA, Aitken JD, Vijay-Kumar M, Gewirtz AT. 2012. Toll-like receptor-gut microbiota interactions: perturb at your own risk! Annual Review of Physiology 74(1):177–198
20. Cash HL, Whitham CV, Behrendt CL, Hooper LV. 2006. Symbiotic bacteria direct expression of an intestinal bactericidal lectin. Science 313(5790):1126–1130
21. Claesson, M. J. et al.2009. Comparative analysis of pyrosequencing and a phylogenetic microarray for exploring microbial community structures in the human distal intestine. PLoS ONE 4, e6669.
22. Clarke G, Stilling RM, Kennedy PJ, Stanton C, Cryan JF, Dinan TG. 2014. Minireview: gut microbiota: the neglected endocrine organ. Molecular Endocrinology 28(8):1221–1238
23. Cockburn DW, Koropatkin NM. 2016. Polysaccharide degradation by the intestinal microbiota and its influence on human health and disease. Journal of Molecular Biology 428(16):3230–3252
24. Collado MC, Rautava S, Aakko J, Isolauri E, Salminen S. 2016. Human gut colonisation may be initiated in utero by distinct microbial communities in the placenta and amniotic fluid. Scientific Reports 6(1):23129

25. Conlon, M. A., and Bird, A. R.2014. The impact of diet and lifestyle on gut microbiota and human health. Nutrients 7, 17–44.
26. Cook, K. L. & Sayler, G. S.2003. Environmental application of array technology: promise, problems and practicalities. Curr. Opin. Biotechnol. 14, 311–318.
27. Costello EK, Stagaman K, Dethlefsen L, Bohannan BJ, Relman DA.2012. The application of ecological theory toward an understanding of the human microbiome. Science.336:1255–1262.
28. Cullen TW, Schofield WB, Barry NA, Putnam EE, Rundell EA, Trent MS, Degnan PH, Booth CJ, Yu H, Goodman AL. 2015. Antimicrobial peptide resistance mediates resilience of prominent gut commensals during inflammation. Science 347(6218):170–175
29. Dalmasso G, Nguyen HTT, Yan Y, Laroui H, Charania MA, Ayyadurai S, Sitaraman SV, Merlin D, DeLeo FR. 2011. Microbiota modulate host gene expression via microRNAs. PLOS ONE 6(4):e19293
30. David LA, Maurice CF, Carmody RN, Gootenberg DB, Button JE, Wolfe BE, Ling AV, Devlin AS, Varma Y, Fischbach MA, Biddinger SB, Dutton RJ, Turnbaugh PJ. 2014. Diet rapidly and reproducibly alters the human gut microbiome. Nature 505(7484):559–563
31. DeAngelis, K. M. et al.2011. PCR amplificationindependent methods for detection of microbial communities by the high-density microarray PhyloChip. Appl. Environ. Microbiol. 77, 6313–6322.
32. Dethlefsen L, Relman DA. 2010. Incomplete recovery and individualized responses of the human distal gut microbiota to repeated antibiotic perturbation. Proceedings of the National Academy of Sciences of the United States of America 108:4554–4561.
33. De Vos, W. M.2009. Mining the microbes—the human microbiome as model. Microb. Biotechnol. 2, 153–154.
34. Dhakan D.B, Maji A,Sharma A.K.,Saxena R., Pulikkan J., Grace T.,Gomez A.,Scaria J., Amato K.R. and Sharma V.K.2019.The unique composition of Indian gut microbiome, gene catalogue, and associated fecal metabolome deciphered using multi-omics approaches. GigaScience ;8:1-20
35. DiGiulio DB, Romero R, Amogan HP, Kusanovic JP, Bik EM, Gotsch F, Kim CJ, Erez O, Edwin S, Relman DA, Fisk NM. 2008. Microbial prevalence, diversity and abundance in amniotic fluid during preterm

labor: a molecular and culture-based investigation. PLOS ONE 3(8):e3056

36. Djuranovic S, Nahvi A, Green R. 2012. miRNA-mediated gene silencing by translational repression followed by mRNA deadenylation and decay. Science 336(6078):237–240
37. Dominguez-Bello MG, Costello EK, Contreras M, Magris M, Hidalgo G, Fierer N, Knight R. 2010. Delivery mode shapes the acquisition and structure of the initial microbiota across multiple body habitats in newborns. Proceedings of the National Academy of Sciences of the United States of America 107(26):11971–11975
38. Eckburg, P. B. et al.2005. Diversity of the human intestinal microbial flora. Science 308, 1635–1638.
39. Fraher, Marianne & O'Toole, Paul & Quigley, Eamonn. 2012. Techniques used to characterize the gut microbiota: A guide for the clinician. Nature reviews. Gastroenterology & hepatology. 9. 312-22. 10.1038/nrgastro.2012.44.
40. Frank DN, Robertson CE, Hamm CM, Kpadeh Z, Zhang T, Chen H, Zhu W, Sartor RB, Boedeker EC, Harpaz N, et al.2011. Disease phenotype and genotype are associated with shifts in intestinal-associated microbiota in inflammatory bowel diseases. Inflamm Bowel Dis.; 17:179–184.
41. Friedrich MJ. 2013. Genomes of microbes inhabiting the body offer clues to human health and disease. JAMA 309(14):1447–1449
42. Ge X, Ding C, Zhao W, Xu L, Tian H, Gong J, Zhu M, Li J, Li N. 2017. Antibiotics-induced depletion of mice microbiota induces changes in host serotonin biosynthesis and intestinal motility. Journal of Translational Medicine 15(1):13
43. Gilbert, J. A., Blaser, M. J., Caporaso, J. G., Jansson, J. K., Lynch, S. V., and Knight, R. 2018. Current understanding of the human microbiome. Nat. Med. 24, 392.
44. Gill SR, et al.2006. Metagenomic analysis of the human distal gut microbiome. Science.312:1355–1359.
45. Goldenberg RL, Hauth JC, Andrews WW. 2000. Intrauterine infection and preterm delivery. New England Journal of Medicine 342(20):1500–1507

46. Goodrich JK, Waters JL, Poole AC, Sutter JL, Koren O, Blekhman R, et al.2014. Human genetics shape the gut microbiome. Cell.;159:789–99.
47. Groer MW, Luciano AA, Dishaw LJ, Ashmeade TL, Miller E, Gilbert JA. 2014. Development of the preterm infant gut microbiome: a research priority. Microbiome 2(1):38
48. Gross L. 2007. Microbes colonize a baby's gut with distinction. PLOS Biology 5(7):e191
49. Guarner, F.; Malagelada J.R.2003. Gut flora in health and disease. Lancet, 361, 512–519.
50. Gupta, V. K., Paul, S., and Dutta, C. 2017. Geography, ethnicity or subsistence-specific variations in human microbiome composition and diversity. Frontiers in microbiology 8:1162.
51. Hall, A. B., Tolonen, A. C., and Xavier, R. J. 2017. Human genetic variation and the gut microbiome in disease. Nat. Rev. Genet. 18, 690.
52. Hamilton MJ, Weingarden AR, Unno T, Khoruts A, Sadowsky MJ.2013. High-throughput DNA sequence analysis reveals stable engraftment of gut microbiota following transplantation of previously frozen fecal bacteria. Gut microbes.4:125–135.
53. Hansen EE, Lozupone CA, Rey FE, Wu M, Guruge JL, Narra A, Goodfellow J, Zaneveld JR, McDonald DT, Goodrich JA, et al.2011. Pan-genome of the dominant human gut-associated archaeon, Methanobrevibacter smithii, studied in twins. Proc Natl Acad Sci U S A.; 108(Suppl 1): 4599–4606.
54. Hasan N, Yang H. 2019. Factors affecting the composition of the gut microbiota, and its modulation. PeerJ 7:e7502
55. Hasler R, Kautz C, Rehman A, et al.2018. The antibiotic resistome and microbiota landscape of refugees from Syria, Iraq and Afghanistan in Germany. Microbiome;6(1):37.
56. Hehemann J-H, Correc G, Barbeyron T, Helbert W, Czjzek M, Michel G. 2010. Transfer of carbohydrate-active enzymes from marine bacteria to Japanese gut microbiota. Nature 464(7290):908–912
57. Heijtz RD, Wang S, Anuar F, Qian Y, Björkholm B, Samuelsson A, Hibberd ML, Forssberg H, Pettersson S.2011. Normal gut microbiota modulates brain development and behavior. Proc. Natl. Acad. Sci. USA.;108:3047–3052.

58. Hermann-Bank, M.L.; Skovgaard, K.; Stockmarr, A.; Larsen, N.; Molbak, L.2013. The gut microbiotassay: A high-throughput qPCR approach combinable with next generation sequencing to study gut microbial diversity. BMC Genomics, 14, doi:10.1186/1471-2164-14-788.
59. Hill, D.A.; Artis, D.2010. Intestinal bacteria and the regulation of immune cell homeostasis. Annu. Rev. Immunol., 28, 623–667.
60. Honda K, Littman DR.2012. The microbiome in infectious disease and inflammation. Annu Rev Immunol.30:759–95.
61. Hooper LV. 2009. Do symbiotic bacteria subvert host immunity? Nature Reviews Microbiology 7(5):367–374
62. Houghteling PD, Walker WA. 2015. Why is initial bacterial colonization of the intestine important to the infant's and child's health? Journal of Pediatric Gastroenterology and Nutrition 60(3):294–307
63. Ichinohe T, Pang IK, Kumamoto Y, Peaper DR, Ho JH, Murray TS, Iwasaki A.2011. Microbiota regulates immune defense against respiratory tract influenza A virus infection. Proc. Natl. Acad. Sci. USA.;108:5354–5359.
64. Imhann,F.,Vila,A.V.,Bonder,M.J.,Fu,J.,Gevers,D.,Visschedijk,M.C., et al. 2018. Interplay of host genetics and gut microbiota underlying the onset and clinical presentation of inflammatory bowel disease. Gut 67, 108–119
65. Ingham, C. J. et al.2007. The micro-Petri dish, a millionwell growth chip for the culture and highthroughput screening of microorganisms. Proc. Natl Acad. Sci. USA 104, 18217–18222.
66. Isaac S, Scher JU, Djukovic A, Jime´nez N, Littman DR, Abramson SB, Pamer EG, Ubeda C. 2016. Short-and long-term effects of oral vancomycin on the human intestinal microbiota. Journal of Antimicrobial Chemotherapy 72(1):128–136
67. Jakobsson HE, Jernberg C, Andersson AF, Sjo¨lund-Karlsson M, Jansson JK, Engstrand L. 2010. Short-term antibiotic treatment has differing long-term impacts on the human throat and gut microbiome. PLOS ONE 5(3):e9836
68. Jalanka-Tuovinen, J.; Salonen, A.; Nikkila, J.; Immonen, O.; Kekkonen, R.; Lahti, L, Palva, A.; de Vos, W.M.2011. Intestinal microbiota in healthy adults: Temporal analysis reveals individual and

common core and relation to intestinal symptoms. PLoS ONE , 6, e23035.

69. Jeffery, I. B. et al.2011. An irritable bowel syndrome subtype defined by species-specific alterations in faecal microbiota. Gut http://dx.doi.org/10.1136/gutjnl-2011-301501.
70. Jernberg C, Lo¨fmark S, Edlund C, Jansson JK. 2007. Long-term ecological impacts of antibiotic administration on the human intestinal microbiota. ISME Journal 1(1):56–66
71. Jime´nez E, Marı´n ML, Martı´n R, Odriozola JM, Olivares M, Xaus J, Ferna´ndez L, Rodrı´guez JM. 2008. Is meconium from healthy newborns actually sterile? Research in Microbiology 159(3):187–193
72. Julia K. Goodrich, Jillian L. Waters, Angela C. Poole, Jessica L. Sutter, Omry Koren, Ran Blekhman, Michelle Beaumont, William Van Treuren, Rob Knight, Jordana T. Bell, Timothy D. Spector, Andrew G. Clark, and Ruth E. Ley.2014. Human genetics shape the gut microbiome. Cell. ; 159(4): 789–799.
73. Kalla R, Ventham NT, Kennedy NA, Quintana JF, Nimmo ER, Buck AH, Satsangi J. 2014. MicroRNAs: new players in IBD. Gut 64(3):504–517
74. Kane M, Case LK, Kopaskie K, Kozlova A, MacDearmid C, Chervonsky AV, Golovkina TV.2011. Successful transmission of a retrovirus depends on the commensal microbiota. Science.334:245–249.
75. Khachatryan ZA, Ktsoyan ZA, Manukyan GP, Kelly D, Ghazaryan KA, Aminov RI.2008. Predominant role of host genetics in controlling the composition of gut microbiota. PLoS ONE. 3:e3064.
76. Klingensmith NJ, Coopersmith CM. 2016. The gut as the motor of multiple organ dysfunction in critical illness. Critical Care Clinics 32(2):203–212
77. Knights D, Costello EK, Knight R.2011. Supervised classification of human microbiota. FEMS Microbiol. Rev.35:343–359.
78. Kostic AD, Gevers D, Pedamallu CS, Michaud M, Duke F, Earl AM, Ojesina AI, Jung J, Bass AJ, Tabernero J, et al.2011. Genomic analysis identifies association of Fusobacterium with colorectal carcinoma. Genome Res.22:292–298.

79. Langdon, A., Crook, N., and Dantas, G.2016. The effects of antibiotics on the microbiome throughout development and alternative approaches for therapeutic modulation. Genome medicine 8, 39.
80. Lange, K., Buerger, M., Stallmach, A., and Bruns, T.2016. Effects of antibiotics on gut microbiota. Dig. Dis. 34, 260–268.
81. Lax, S., Smith, D. P., Hampton-Marcell, J., Owens, S. M., Handley, K. M., Scott, N. M., et al. 2014. Longitudinal analysis of microbial interaction between humans and the indoor environment. Science 345, 1048–1052
82. LeBlanc JG, Milani C, de Giori GS, Sesma F, van Sinderen D, Ventura M.2013. Bacteria as vitamin suppliers to their host: a gut microbiota perspective. Curr Opin Biotech;24:160–8.
83. Lee Y-M, Kim K-S, Jacobs DR Jr, Lee D-H. 2017. Persistent organic pollutants in adipose tissue should be considered in obesity research. Obesity Reviews 18(2):129–139
84. Lehrer, R. I. & Lu, W.2012. α-Defensins in human innate immunity. Immunol. Rev. 245, 84–112.
85. Levin, B. J. et al.2017. A prominent glycyl radical enzyme in human gut microbiomes metabolizes trans-4-hydroxyl-proline. Science 355, eaai8386.
86. Lewis BB, Buffie CG, Carter RA, Leiner I, Toussaint NC, Miller LC, Gobourne A, Ling L, Pamer EG. 2015. Loss of microbiota-mediated colonization resistance to Clostridium difficile infection with oral vancomycincomparedwithmetronidazole.JournalofInfectiousDiseases 212(10):1656–1665
87. Ley, R. E. et al.2008. Evolution of mammals and their gut microbes. Science 320, 1647–1651.
88. Li, F., Hullar, M. A. & Lampe, J. W.2007. Optimization of terminal restriction fragment polymorphism (TRFLP) analysis of human gut microbiota. J. Microbiol. Methods 68, 303–311.
89. Link A, Becker V, Goel A, Wex T, Malfertheiner P, Hoheisel JD. 2012. Feasibility of fecal microRNAs as novel biomarkers for pancreatic cancer. PLOS ONE 7(8):e42933
90. Liu S, Da Cunha AP, Rezende RM, Cialic R, Wei Z, Bry L, Comstock LE, Gandhi R, Weiner HL. 2016. The host shapes the gut microbiota via fecal microRNA. Cell Host & Microbe 19(1):32–43

91. Lloyd-Price, J., Abu-Ali, G., and Huttenhower, C.2016. The healthy human microbiome. Genome medicine 8, 51.
92. Lozupone, C. A., Stombaugh, J. I., Gordon, J. I., Jansson, J. K., and Knight, R.2012. Diversity, stability and resilience of the human gut microbiota. Nature 489, 220
93. Macfarlane, G.T.; Cummings, J.H.1999. Probiotics and prebiotics: Can regulating the activities of intestinal bacteria benefit health? BMJ, *318*, 999–1003.
94. Macfarlane, S.; Steed, H.; Macfarlane, G.T.2009. Intestinal bacteria and inflammatory bowel disease. Crit. Rev. Clin. Lab. Sci., *46*, 25–54.
95. Macpherson AJ, Uhr T. 2004. Induction of protective IgA by intestinal dendritic cells carrying commensal bacteria. Science 303(5664):1662–1665
96. Maji A, Misra R, Dhakan DB, et al.2018. Gut microbiome contributes to impairment of immunity in pulmonary tuberculosis patients by alteration of butyrate and propionate producers. Environ Microbiol;20(1):402–19.
97. Manichanh, C. et al.2006. Reduced diversity of faecal microbiota in Crohn's disease revealed by a metagenomic approach. Gut 55, 205–211.
98. Mariat, D. et al.2009. The Firmicutes/Bacteroidetes ratio of the human microbiota changes with age. BMC Microbiol. 9, 123.
99. Marsh, T. L.1999. Terminal restriction fragment length polymorphism (T-RFLP): an emerging method for characterizing diversity among homologous populations of amplification products. Curr. Opin. Microbiol. 2, 323–327.
100. Maslowski KM, Mackay CR.2011. Diet, gut microbiota and immune responses. Nat Immunol.12:5–9.
101. Massacand JC, Kaiser P, Ernst B, Tardivel A, Bu¨rki K, Schneider P, Harris NL. 2008. Intestinal bacteria condition dendritic cells to promote IgA production. PLOS ONE 3(7):e2588
102. Matamoros S, Gras-Leguen C, Le Vacon F, Potel G, de La, Cochetiere MF.2013. Development of intestinal microbiota in infants and its impact on health. Trends Microbiol.21:167–73.
103. McCartney, A. L.2002. Application of molecular biological methods for studying probiotics and the gut flora. Br. J. Nutr. 88 (Suppl. 1), S29–S37.

104. McKnite AM, Perez-Munoz ME, Lu L, Williams EG, Brewer S, Andreux PA, et al.2012. Murine gut microbiota is defined by host genetics and modulates variation of metabolic traits. PLoS One. 7:e39191.
105. Mohan V, Sandeep S, Deepa R, et al.2007. Epidemiology of type 2 diabetes: Indian scenario. Indian J Med Res ;125(3):217– 30.
106. Morrison DJ, Preston T.2016. Formation of short chain fatty acids by the gut microbiota and their impact on human metabolism. Gut Microbes.7:189–200.
107. M'Rabet L, Vos AP, Boehm G, Garssen J. 2008. Breast-feeding and its role in early development of the immune system in infants: consequences for health later in life. Journal of Nutrition 138(9):1782S–1790S.
108. Mukherjee S, Zheng H, Derebe MG, Callenberg KM, Partch CL, Rollins D, et al. 2014. Antibacterial membrane attack by a pore-forming intestinal C-type lectin. Nature 505(7481):103–107.
109. Mueller, C.; Macpherson, A.J. 2006. Layers of mutualism with commensal bacteria protect us from intestinal inflammation. Gut, *55*, 276–284.
110. Muyzer, G., de Waal, E. C. & Uitterlinden, A. G.1993. Profiling of complex microbial populations by denaturing gradient gel electrophoresis analysis of polymerase chain reaction-amplified genes coding for 16S rRNA. Appl. Environ. Microbiol. 59, 695–700.
111. Muyzer, G.1999. DGGE/TGGE a method for identifying genes from natural ecosystems. Curr. Opin. Microbiol. 2, 317–322.
112. Nagpal R, Tsuji H, Takahashi T, Nomoto K, Kawashima K, Nagata S, Yamashiro Y. 2017. Ontogenesis of the gut microbiota composition in healthy, full-term, vaginally born and breastfed infants over the first 3 years of life: a quantitative bird's-eye view. Frontiers in Microbiology 8:1388
113. Neu J.2016. The microbiome during pregnancy and early postnatal life. Semin Fetal Neonatal Med.;21:373–9.
114. Noor, S. O. et al.2010 Ulcerative colitis and irritable bowel patients exhibit distinct abnormalities of the gut microbiota. BMC Gastroenterol. 10, 134.

115. Ouwehand A, Isolauri E, Salminen S. 2002. The role of the intestinal microflora for the development of the immune system in early childhood. European Journal of Nutrition 41:i32–i37

116. Osborn, A. M., Moore, E. R. & Timmis, K. N. 2000. An evaluation of terminal-restriction fragment length polymorphism (T-RFLP) analysis for the study of microbial community structure and dynamics. Environ. Microbiol. 2, 39–50.

117. Paliy, O., Kenche, H., Abernathy, F. & Michail, S. 2009. High-throughput quantitative analysis of the human intestinal microbiota with a phylogenetic microarray. Appl. Environ. Microbiol. 75, 3572–3579.

118. Palmer, C. et al. 2006. Rapid quantitative profiling of complex microbial populations. Nucleic Acids Res. 34, e5.

119. Palmer, C., Bik, E. M., Digiulio, D. B., Relman, D. A. & Brown, P. O. 2007. Development of the human infant intestinal microbiota. PLoS Biol. 5, e177

120. Pedersen HK, Gudmundsdottir V, Nielsen HB, et al. 2016. Human gut microbes impact host serum metabolome and insulin sensitivity. Nature;535(7612):376–81.

121. Peixin Fan, Beilei Bian, Lin Teng, Corwin D. Nelson,J. Driver, Mauricio A. Elzo,Kwangcheol C. Jeong.2020. Host genetic effects upon the early gut microbiota in a bovine model with graduated spectrum of genetic variation. The ISME Journal 14:302–317.

122. Petar Scepanovic, Flavia Hodel, Stanislas Mondot, Valentin Partula, Allyson Byrd, Christian Hammer, et al. 2019. A comprehensive assessment of demographic, environmental, and host genetic associations with gut microbiome diversity in healthy individuals. Microbiome. 7:130.

123. Peterson, J. et al. 2009. The NIH Human Microbiome Project. Genome Res. 19, 2317–2323.

124. Ponnusamy, K., Choi, J. N., Kim, J., Lee, S. Y. & Lee, C. H.2011. Microbial community and metabolomic comparison of irritable bowel syndrome faeces. J. Med. Microbiol. 60, 817–827.

125. Pulikkan J, Maji A, Dhakan DB, et al. 2018. Gut microbial dysbiosis in Indian children with autism spectrum disorders. Microb Ecol;76(4):1102–14.